THE ORGANIC CHEMISTRY OF DRUG SYNTHESIS

Volume 7

DANIEL LEDNICER
North Bethesda, MD

BICENTENNIAL
1807
WILEY
2007
BICENTENNIAL

WILEY-INTERSCIENCE
A JOHN WILEY & SONS, INC., PUBLICATION

Published by John Wiley & Sons, Inc., Hoboken, New Jersey
Published simultaneously in Canada

For general information on our other products and services or for technical support, please contact our Customer Care Department within the United States at (800) 762-2974, outside the United States at (317) 572-3993 or fax (317) 572-4002.

Wiley also publishes it books in variety of electronic formats. Some content that appears in print may not be available in electronic formats. For more information about Wiley products, visit our web site at www.wiley.com.

Wiley Bicentennial Logo: Richard J. Pacifico

Library of Congress Cataloging-in-publication Data is available.

ISBN 978-0-470-10750-8

Printed in the United States of America

10 9 8 7 6 5 4 3 2 1

To the memory of I. Moyer Hunsberger and Melvin S. Newman
who set me on course...

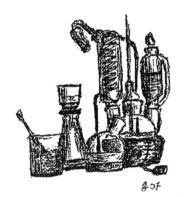

CONTENTS

PREFACE

The first volume of *The Organic Chemistry of Drug Synthesis* was originally visualized as a single free-standing book that outlined the syntheses of most drugs that had been assigned non-proprietary names in 1975 at the time the book was written. Within a year or so of publication in 1977, it had become evident that a good many drugs had been overlooked. That and the encouraging reception of the original book led to the preparation of a second volume. That second book not only made up for the lacunae in the original volume but also covered additional new drug entities as well. With that second volume assignment of non-proprietary names by USAN became the criterion for inclusion. That book, published in 1980, thus included in addition all agents that had been granted USAN since 1976. What had been conceived as a single book at this point became a series. The roughly 200 new USAN coined every five years over the next few decades turned out to nicely fit a new volume in the series. This then dictated the frequency for issuing new compendia. After the most recent book in the series, *Volume 6*, was published in 1999, it became apparent that a real decline in the number of new drug entities assigned non-proprietary names had set in. The customary half-decade interval between books was apparently no longer appropriate.

A detailed examination of the 2005 edition of the *USAN Dictionary of Drug Names* turned up 220 new non-proprietary names that had been assigned since the appearance of Volume 6. Many of these compounds represent quite novel structural types first identified by sophisticated new cell-based assays. This clearly indicated the need for the present volume in the series *The Organic Chemistry of Drug Synthesis.*

This new book follows the same format as the preceding volumes. Compounds are classed by their chemical structures rather than by their biological activities. This is occasionally awkward since compounds with the same biological activity but significantly different structures are relegated to different chapters, a circumstance particularly evident with estrogen antagonists that appear in three different chapters. The cross index found at the end of the book, it is hoped, partly overcomes this problem. The syntheses are discussed from an organic chemist's point of view, accompanied by the liberal use of flow diagrams. As was the case in the preceding volumes, a thumbnail explanation of the biological activity of each new compound precedes the discussion of its biological activity.

Several trends in the direction of drug discovery research seemed to emerge during the preparation of this book. Most of the preceding volumes included one or more therapeutic classes populated by many structurally related potential drugs. *Volume 6* for example described no fewer than a half dozen HIV-protease inhibitors and a similar number of the "triptan" drugs aimed at treating migraine. The distribution of therapeutic activities in the present volume is quite distinct from that found in the earlier books. This new set, for example, includes a sizeable number of antineoplastic and antiviral agents. These two categories together in fact account for just over one third of the compounds in the present volume. The antitumor candidates are further distinct in that specific agents act against very specific tumor-related biological end points. This circumstance combined with mechanism based design in other disease areas probably reflect the widespread adoption of in-vitro screening in the majority of pharmaceutical research laboratories.

The use of combinatorial chemistry for generating libraries to feed in-vitro screens has also become very prevalent over the past decade. This book is silent on that topic since compounds are only included when in a quite advanced developmental stage. Some of the structures that include strings of unlikely moieties suggest that those compounds may have been originally prepared by some combinatorial process.

The internet has played a major role in finding the articles and patents that were required to put this account together. The NIH-based website PubChem was an essential resource for finding structures of compounds that appear in this book; hits more often than not include CAS Registry Numbers. References to papers on the synthesis of compounds could sometimes be found with the other NIH source PubMed. The ubiquitous Google was also quite helpful for finding sources for syntheses. In some of the earlier volumes, references to patents were accompanied by references to the corresponding CAS abstract since it was often difficult to access patents. The availability of actual images of all patents from either the U.S. patent office (www.uspto.gov) or those from European elsewhere (http://ep.espacenet.com) has turned the situation around. There was always the rather pricey STN online when all else failed.

This volume, like its predecessors, is aimed at practicing medicinal and organic chemists as well as graduate and advanced undergraduate students in organic and medicinal chemistry. The book assumes a good working knowledge of synthetic organic chemistry and some exposure to modern biology.

As a final note, I would like to express my appreciation to the staff of the library in Building 10 of the National Institutes of Health. Not only were they friendly and courteous but they went overboard in fulfilling requests that went well beyond their job descriptions.

CHAPTER 1

OPEN-CHAIN COMPOUNDS

Carbocyclic or heterocyclic ring systems comprise the core of chemical structures of the vast majority of therapeutic agents. This finding results in the majority of drugs exerting their effect by their actions at receptor or receptor-like sites on cells, enzymes, or related entities. These interactions depend on the receiving site being presented with a molecule that has a well-defined shape, distribution of electron density, and array of ionic or ionizable sites, which complement features on the receptor. These requirements are readily met by the relatively rigid carbocyclic or heterocyclic molecules. A number of important drugs cannot, however, be assigned to one of those structural categories. Most of these agents act as false substrates for enzymes that handle peptides. The central structural feature of these compounds is an open-chain sequence that mimics a corresponding feature in the normal peptide. Although these drugs often contain carbocyclic or heterocyclic rings in their structures, these features are peripheral to their mode of action. Chapter 1 concludes with a few compounds that act by miscellany and mechanisms and whose structures do not fit other classifications.

The Organic Chemistry of Drug Synthesis, Volume 7. By Daniel Lednicer
Copyright © 2008 John Wiley & Sons, Inc.

1. PEPTIDOMIMETIC COMPOUNDS

A. Antiviral Protease Inhibitors

1. Human Immunodeficiency Virus. The recognition of acquired immune deficiency syndrome (AIDS) in the early 1980s and the subsequent explosion of what had seemed at first to be a relatively rare disease into a major worldwide epidemic, lent renewed emphasis to the study of virus-caused disease. Treatment of viral disease is made particularly difficult by the fact that the causative organism, the virion, does not in the exact meaning of the word, replicate. Instead, it captures the reproductive mechanism of infected cells and causes those to produce more virions. Antiviral therapy thus relies on seeking out processes that are vital for producing those new infective particles. The first drugs for treating human immunodeficiency virus (HIV) infection comprised heterocyclic bases that interfered with viral replication by interrupting the transcription of viral ribonucleic acid (RNA) into the deoxyribonucleic acid (DNA) required by the host cell for production of new virions. The relatively fast development of viral strains resistant to these compounds has proven to be a major drawback to the use of these reverse transcriptase inhibitors. The drugs do, however, still form an important constituent in the so-called cocktails used to treat AIDS patients. Some current reverse transcriptase inhibitors are described in Chapters 4 and 6. The intense focus on the HIV virus revealed yet another point at which the disease may be tackled. Like most viruses, HIV comprises a packet of genetic material, in this case RNA, encased in a protein coat. This protein coat provides not only protection from the environment, but also includes peptides that recognize features on host cells that cause the virion to bind to the cell and a few enzymes crucial for replication. Many normal physiological peptides are often elaborated as a part of a much larger protein. Specialized peptidase enzymes are required to cut out the relevant protein. This proved to be the case with the peptides required for forming the envelopes for newly generated virions. Compounds that inhibit the scission of the protein elaborated by the infected host, the HIV protease inhibitors, have provided a valuable set of drugs for treatment of infected patients. The synthesis of four of those drugs were outlined in Volume 6 of this series. Work on compounds in this class has continued apace as evidenced by the half dozen new protease inhibitors that have been granted nonproprietary names since then.

As noted in Volume 6, the development of these agents was greatly facilitated by a discovery in a seemingly unrelated area. Research aimed at development of renin inhibitors as potential antihypertensive agents had led to the discovery of compounds that blocked the action of this peptide cleaving enzyme. The amino acid sequence cleaved by renin was found to be fortuitously the same as that required to produce the HIV peptide coat. Structure–activity studies on renin inhibitors proved to be of great value for developing HIV protease inhibitors. Incorporation of an amino alcohol moiety proved crucial to inhibitory activity for many of these agents. This unit is closely related to the one found in the statine, an unusual amino acid that forms part of the pepstatin, a fermentation product that inhibits protease enzymes.

Statine   Transition State   Protease Inhibitor
                                               Fragment

This moiety may be viewed as a carbon analogue of the transition state in peptide cleavage. The fragment is apparently close enough in structure to such an intermediate as to fit the cleavage site in peptidase enzymes. Once bound, this inactivates the enzyme as it lacks the scissile carbon–nitrogen bond. All five newer HIV protease inhibitors incorporate this structural unit.

One scheme for preparing a key intermediate for incorporating that fragment begins with the chloromethyl ketone (**1**) derived from phenyl-alanine, in which the amine is protected as a carbobenzyloxy (Cbz) group. Reduction of the carbonyl group by means of borohydride affords a mixture of aminoalcohols. The major syn isomer **2** is then iso-lated. Treatment of **2** with base leads to internal displacement of halogen and formation of the epoxide (**3**).[1]

1   2   3

The corresponding analogue (**4**) in which the amine is protected as a *tert*-butyloxycarbonyl function (*t*-BOC) comprises the starting material for the HIV protease inhibitor **amprenavir** (**12**). Reaction of **4** with isobutyl amine leads to ring opening of the oxirane and formation of the aminoalcohol (**5**). The thus-formed secondary amine in the product is converted to the sulfonamide (**6**) by exposure to *p*-nitrobenzenesulfonyl chloride. The *t*-BOC protecting group is then removed by exposure to acid leading to the primary amine (**10**). In a convergent scheme, chiral 3-hydroxytetrahydrofuran (**8**) is allowed to react with bis(*N*-succinimidooxy)carbonate (**7**). The hydroxyl displaces one of the *N*-hydroxysuccinimide groups to afford the tetrahydrofuran (THF) derivative (**9**) equipped with a highly activated leaving group. Reaction of this intermediate with amine 10 leads to displacement of the remaining *N*-hydroxysuccinimide and incorporation of the tetrahydrofuryl moiety as a urethane (**11**). Reduction of the nitro group then affords the protease inhibitor (**12**).[2]

Much the same sequence leads to a protease inhibitor that incorporates a somewhat more complex furyl function-linked oxygen heterocyclic. This fused bis(tetrahydrofuryl) alcohol (**16**) was designed to better interact with a pocket on the viral protease. The first step in preparing this intermediate consists of reaction of dihydrofuran (**13**) with propargyl alcohol and iodosuccinimide to afford the iodoether (**14**). Free radical displacement of the iodine catalyzed by cobaloxime leads to the fused perhydrofuranofuran (**15**). The exomethylene group in the product is then cleaved by means of ozone; reductive workup of the ozonide leads to racemic **16**. The optically pure single entity (**17**) is then obtained by resolution of the initial mixture of isomers with immobilized lipase.[3]

That product (**17**) is then converted to the activated N-hydoxysuccinimide derivative **18** as in the case of the monocyclic furan. Reaction with the primary amine **10** used to prepare amprenavir then leads to the urethane (**19**). Reduction of the nitro group then affords **darunavir**[4] (**20**).

The synthesis of the amprenavir derivative, which is equipped with a solubilizing phosphate group, takes a slightly different course from that used for the prototype. The protected intermediate **5** used in the synthesis of **12** is allowed to react with benzyloxycarbonyl chloride to provide the doubly protected derivative **21**, a compound that bears a *t*-BOC group on one nitrogen and a Cbz grouping on the other. Exposure to acid serves to remove the *t*-BOC group, affording the primary amine **22**. This compound is then condensed with the activated intermediate **9** used in the preparation of the prototype to yield the urethane **23**. Catalytic hydrogenation then removes the remaining protecting group to give the secondary amine **24**. Reaction as before with *p*-nitrobenzenesulfonyl chloride gives the sulfonamide **25**. This intermediate is allowed to react with phosphorus oxychloride under carefully controlled conditions. Treatment with aqueous acid followed by a second catalytic hydrogenation affords the water soluble protease inhibitor **fosamprenavir (26)**.[5]

The preceding three antiviral agents tend to differ form each other by only relatively small structural details. The next protease inhibitor includes some significant structural differences though it shares a similar central aminoalcohol sequence that is presumably responsible for its activity. Construction of one end of the molecule begins with protection of the carbonyl function in *p*-bromobenzaldehyde (**27**) as its methyl acetal (**28**) by treatment with methanol in the presence of acid. Reaction of that intermediate with the Grignard reagent from 4-bromopyridine leads to unusual displacement of bromine from the protected benzaldehyde and formation of the coupling product. Mild aqueous acid restores the aldehyde function to afford **29**. This compound is then condensed with carbethoxy hydrazine to form the respective hydrazone; reduction of the imine function leads to the substituted hydrazine (**30**). Reaction of **30** with the by-now familiar amino-epoxide (**4**) results in oxirane opening by attack of the basic nitrogen in the hydrazine (**30**) and consequent formation of the addition product **31**. The *t*-BOC protecting group is then removed by treatment with acid. The final step comprises acylation of the free primary amine in **32** with the acid chloride from the *O*-methyl urethane (**33**). This last compound (**32**) is a protected version of an unnatural α-aminoacid that can be viewed as valine in with an additional methyl group on what had been the side-chain secondary carbon atom. Thus, the protease inhibitor **atazanavir** (**34**) is obtained.[6]

A terminal cyclic urea derivative of valine is present at one terminus in **lopinavir** (**43**). Preparation of this heterocyclic moiety begins with conversion of valine (**35**) to its phenoxycarbonyl derivative by reaction with the corresponding acid chloride. Alkylation of the amide nitrogen with 3-chloropropylamine in the presence of base under very carefully controlled pH results in displacement of the phenoxide group to give the urea intermediate (**37**). This compound then spontaneously undergoes internal displacement of chlorine to form the desired derivative (**38**).

The statine-like aminoalcohol function in this compound differs from previous examples by the presence of an additional pendant benzyl group; the supporting carbon chain is of necessity longer by one member. Condensation of that diamine (**39**),[7] protected at one end as its *N,N*-dibenzyl derivative, with 2,6-dimethylphenoxyacetic acid (**40**) gives the corresponding amide (**41**). Hydrogenolysis then removes the benzyl protecting groups to afford primary amine **42**. Condensation of that with intermediate **34** affords the HIV protease inhibitor **43**.[8]

2. Human Rhinovirus. Human rhinoviruses are one of the most frequent causes of that affliction that accompanies cooling weather, the common cold. This virus also consists of a small strand of RNA enveloped in a peptide coat. Expression of fresh virions in this case depends on provision of the proper peptide by the infected host cell. That in turn hinges on excision of that peptide from the larger initially produced protein. Protease inhibitors have thus been investigated as drugs for treating rhinovirus infections. The statine-based HIV drugs act by occupying the scission site of the protease enzyme and consequently preventing access by the HIV-related substrate. That binding is, however, reversible in the absence of the formation of a covalent bond between drug and enzyme. A different strategy was employed in the research that led to the rhinovirus protease inhibitor **rupinavir** (**58**). The molecule as a whole is again designed to fit the protease enzyme, as in the case of the anti-HIV compounds. In contrast to the latter compound, however, this agent incorporates a moiety that will form a covalent bond with the enzyme, in effect inactivating it with finality. The evocative term "suicide inhibitor" has sometimes been used for this approach since both the substrate and drug are destroyed.

The main part of the somewhat lengthy convergent synthesis consists of the construction of the fragment that will form the covalent bond with the enzyme. The unsaturated ester in this moiety was designed to act as a Michael acceptor for a thiol group on a cysteine residue known to be present at the active site. The preparation of that key fragment starts with the protected form of chiral 3-amino-4-hydroxybutyric acid (**44**); note that the oxazolidine protecting group simply comprises a cyclic hemiaminal of the aminoalcohol with acetone. The first step involves incorporation of a chiral auxiliary to guide introduction of an additional carbon atom. The carboxylic acid is thus converted to the corresponding acid chloride and that reacted with the (*S*)-isomer of the by-now classic oxazolidinone (**45**) to give derivative **46**. Alkylation of the enolate from **46** with allyl iodide gives the corresponding allyl derivative (**47**) as a single enantiomer. The double bond is then cleaved with ozone; reductive workup of the ozonide affords the aldehyde (**48**). Reductive amination of the carbonyl group with 2,6-dimethoxybenzylamine in the presence of cyanoborohydride proceeds to the corresponding amine **49**. This last step in effect introduced a protected primary amino group at that position. The chiral auxiliary grouping is next removed by mild hydrolysis. The initially formed amino acid (**50**) then cyclizes to give the five-membered lactam (**51**). Treatment under stronger hydrolytic conditions subsequently serves to open the cyclic hemiaminal grouping to reveal the free aminoalcohol

(**52**). Swern-type oxidation of the terminal hydroxyl group in this last intermediate affords an intermediate (**53**) that now incorporates the aldehyde group required for building the Michael acceptor function. Thus reaction of that compound with the ylide from ethyl 2-diethoxyphosphonoacetate adds two carbon atoms and yields the acrylic ester (**54**).

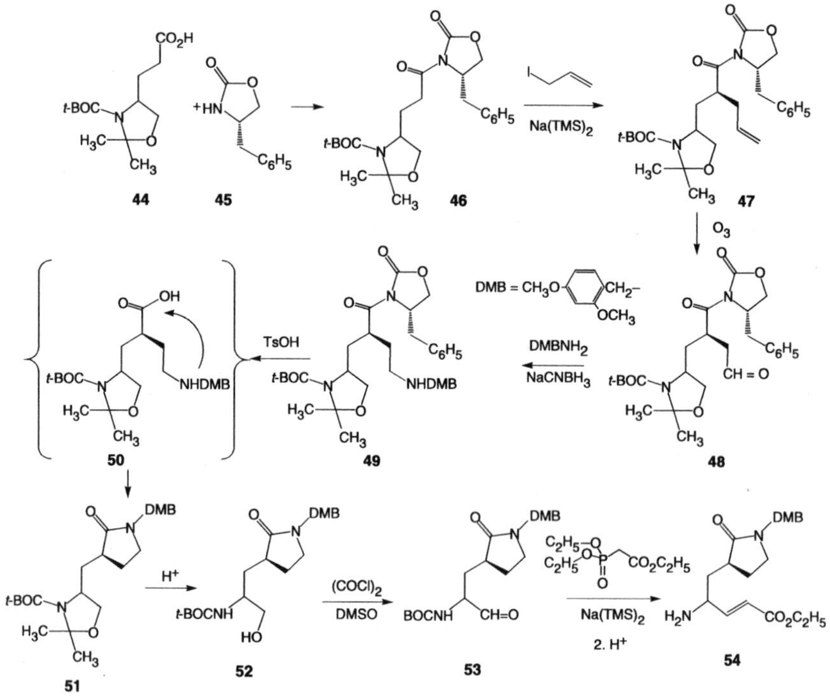

TMS = tetramethylsilane
DMB = 2,4-dimethoxybenzyl
DMSO = dimethyl sulfoxide
Ts = Tosyl

The remaining portion of the molecule is prepared by the condensation of *N*-carbobenzyloxyleucine with *p*-fluorophenylalanine to yield the protected dipeptide (**55**). Condensation of that intermediate with the Michael acceptor fragment (**54**) under standard peptide-forming conditions leads to the tripeptide-like compound (**53**). Reaction of **53** with dichlorodicyanoquinone (DDQ) leads to unmasking of the primary amino group at the end of the chain by oxidative loss of the DMB protecting group. Acylation of that function with isoxazole (**55**) finally affords the rhinovirus protease inhibitor rupinavir (**58**).[9]

2. MISCELLANEOUS PEPTIDOMIMETIC COMPOUNDS

Polymers of the peptide tubulin make up the microtubules that form the microskeleton of cells. Additionally, during cell division these filaments pull apart the nascent newly formed pair of nuclei. Compounds that interfere with tubulin function and thus block this process have provided some valuable antitumor compounds. The vinca alkaloid drugs vincristine and vinblastine, for example, block the self-assembly of tubulin into those filaments. Paclitaxel, more familiarly known as Taxol, interestingly stabilizes tubulin and in effect freezes cells into mid-division. Screening of marine natural products uncovered the cytotoxic tripeptide-like compound hemiasterlin, which owed its activity to inhibition of tubulin. A synthetic program based on that lead led to the identification of **taltobulin (69)**, an antitumor compound composed, like its model, of sterically crowded aminoacid analogues. The presence of the nucleophile-accepting acrylate moiety recalls **58**.

One arm of the convergent synthesis begins with the construction of that acrylate-containing moiety. Thus, condensation of the *t*-BOC protected α-aminoaldehyde derived from valine with the carbethoxymethylene phosporane (**60**) gives the corresponding chain extended amino ester (**61**). Exposure to acid serves to remove the protecting group to reveal the primary amine (**62**). Condensation of that intermediate with the tertiary butyl-substituted aminoacid **33**, used in a previous example leads to the protected amide (**63**); the *t*-BOC group in this is again removed with acid unmasking the primary amino group in **64**. Construction of the other major fragment first involves addition of a pair of methyl groups to the benzylic position of pyruvate (**65**). This transform is accomplished under surprisingly mild conditions by simply treating the ketoacid with methyl iodide in the presence of hydroxide. Treatment of product **66** with methylamine and diborane results in reductive amination of the carbonyl group, and thus formation of α-aminoacid **67** as a mixture of the two isomers. Condensation of that with the dipeptide-like moiety **64** under standard peptide-forming conditions gives the amide **68** as a mixture of diastereomers. The isomers are then separated by chromatography; saponification of the terminal ester function of the desired (*SSS*)-isomer affords the antitumor agent taltobulin (**69**).[10]

The alkylating agent cyclophosphamide is one of the oldest U.S. Food and Drug Administration (FDA)-approved antitumor agents, having been in use in the clinic for well over four decades. Though this chemotherapeutic agent is reasonably effective, it is not very selective. The drug affects many sites and is thus very poorly tolerated. Over the years, there has been much research devoted to devising more site-selective related compounds. It was established that a heterocyclic ring in this compound is opened metabolically and then discarded. The active alkylating metabolite comprises the relatively small molecule commonly known as the "phosphoramide mustard".

Cyclophosphamide   Phosphoramide Mustard   Glutathione

This result opens the possibility of delivering this active fragment or a related alkylating function in a large molecule that would itself be recognized by an enzyme involved in cancer progression. As an example, it was observed that many types of cancer tissues often have elevated levels of glutathione transferase, the agent that removes glutathione. A version of the modified natural substrate, glutathione, which carries a phosphoramide alkylating function, has shown activity on various cancers. Reaction of bromoethanol with phosphorus oxychloride affords intermediate **70**. This compound reacts without purification with bis-2-chloroethylamine to give the phosphoramide (**71**), which is equipped with two sets of alkylating groups. Compound **71** is then reacted with the glutathione analogue **72**, in which phenylglycine replaces the glycine residue normally at that position. The bromine atom in intermediate **71** is apparently sufficiently more reactive than the chlorines in the mustards so that displacement by sulfur preferentially proceeds to **73**. Oxidation of the sulfide with hydrogen peroxide affords **canfosfamide (74)**.[11,12]

The D(R) isomer of the amino acid N-methyl-D-aspartate, more commonly known as NMDA serves as the endogenous agonist at a number of central nervous system (CNS) receptor sites. This agent is not only involved in neurotransmission, but also modulates responses elicited by other neurochemicals. A relatively simple peptide-like molecule has been found to act as an antagonist at NMDA receptors. This activity is manifested *in vivo* as antiepileptic activity. This agent in addition blocks the nerve pain suffered by many diabetics, which is often called neuropathic pain. The synthesis begins by protecting the unnatural D-serine

(**75**) as its carbobenzyloxy derivative **76**. This is accomplished by reacting **75** with the corresponding acid chloride. Reaction of the product with methyl iodide in the presence of silver oxide alkylates both the free hydroxyl and the carboxylic acid to give the ether ester (**77**). Saponification followed by coupling with benzylamine leads to the benzylamide (**78**). Hydrogenolysis of the Cbz protecting group (**79**) followed by acylation with acetic anhydride affords **lacosamide (80)**.[13]

As noted in the discussion of canfosfamide, alkylating agents have a long history as a class of compounds used in chemotherapy. The trend is to attach the active electrophillic groups to molecules that will deliver them to specific sites. A simple alkylating agent, **cloretazine (83)**, is being actively pursued because of its promising antitumor activity. Exhaustive methanesulfonation of hydroxyethyl hydrazine with methanesulfonyl chloride yields the *N,N,O*-trimesylate (**81**). Reaction of this intermediate with lithium chloride leads to displacement of the *O*-mesylate by chlorine and formation of the alkylating group in **82**. Treatment of **82** with the notorious methylisocyanate (MCI) yields the antitumor agent cloretazine (**83**).[14,15]

The relatively simple homologue of taurine, 3-aminosulfonic acid (**84a**), also known as homotaurine, is an antagonist of the neurochemical gamma-aminobutyric acid (GABA). Homotaurine has been found to suppress alcoholism in various animal models. Speculation is that this occurs because of its activity against GABA to which it bears a some resemblance. The calcium salt (**84b**) of the N-acetyl derivative has been used to help alcoholics maintain abstinence from alcohol by preventing relapse. The compound is prepared straightforwardly by acylation of homotaurine in the presence of calcium hydroxide and acetic anhydride.[16] The product, **acamprosate calcium (84b)**, was approved by FDA for use in the United States in 2004.

A relatively simple derivative of phenylalanine shows hypoglycemic activity. This compound, **nateglinide**, is usually prescribed for use as an adjunct to either metformin, or one of the thiazolidine hypoglycemic agents. Catalytic reduction of the benzoic acid (**85**) leads to the corresponding substituted cyclohexane as a mixture of isomers. This compound is then esterified with methanol to give the methyl esters (**86**). Treatment with sodium hydride leads **86** to equilibrate to the more stable trans isomer **87** via its enolate. Condensation of **87** with the ester of phenylalanine (**88**) yields nateglinide (**89**) after saponifications.[17]

The hypoglycemic agent **repaglinide** may loosely be classed as a peptidomimetic agent, because it essentially shows the same activity as nateglinide. The actual synthetic route is difficult to decipher from the patent in which it is described. No description is provided for the origin of the starting materials. It is speculated that condensation of the protected monobenzyl ester (**90**) with diamine **91** would lead to the amide (**92**). Hydrogenolysis of the benzyl ester in the product would afford the free acid. Thus, repaglinide (**93**) would be obtained.[18]

90   91

92; R = CH₂C₅H₅
93; R = H

Formation of blood clots is the natural process that preserves the integrity of the circulatory system. Damage to the vasculature sets off an intricate cascade of reactions. These reactions culminate in the formation of a fibrin clot that seals the damaged area preventing the further loss of blood. Surgery, heart attacks, and other traumatic events lead to inappropriate formation of clots that can result in injury by blocking the blood supply to organs and other vital centers. The drugs that have traditionally been used to prevent formation of clots, coumadin and heparin have a very narrow therapeutic ratio, necessitating close monitoring of blood levels of these drugs in patients. One of the first steps in the formation of a clot involves the binding of fibrinogen to specific receptors on the platelets that start the process. A number of fibrinogen inhibitors have recently been developed whose structure is based on the sequence of amino acids in the natural product. Two more recent compounds, **melagartan**, and **xymelagartan**, both contain the amidine (or guanidine) groups that are intended to mimic the similar function in fibrinogen and that characterize this class of drugs.[19]

The synthesis of these agents begins with the hydrogenation of phenylglycine *t*-BOC amide (**94**) to the corresponding cyclohexyl derivative **95**. The free carboxyl group is then coupled with the azetidine (**96**) to afford the amide (**97**). Saponification with lithium hydroxide yields the free acid (**98**). The carboxyl group in that product is then coupled with the benzylamine (**99**), where the amidine group at the para position is protected as the benzyloxycarbonyl derivative to give intermediate **100**. The protecting group on the terminal amino group is then removed by hydrolysis with acid (**101**). The primary amine in this last intermediate is then alkylated with benzyl bromoacetate. Hydrogenolysis removes the protecting groups on the terminal functions in this molecule to afford melagartan (**102**).[20]

Intermediate **100** serves as the starting material for the structurally closely related fibrinogen inhibitor xymelagartan. Hydrogenation over palladium on charcoal removes the protecting group on the amidine function (**103**). This compound is then allowed to react with what is in effect and unusual complex ester of carbonic acid (**104**). The basic nitrogen on the amidine displaces nitrophenol, a good leaving group to afford **106**. Regiochemistry is probably dictated by the greater basicity of the amidine group compared to the primary amine at the other end of the molecule. The amine is then alkylated with the trifluoromethyl-sulfonyl derivative of ethyl hydroxyacetate. Reaction of this last intermediate (**107**) with hydroxylamine result in an exchange of the substitutent on the amidine nitrogen to form an N-hydroxyamidine. Thus, **xymelagartan** (**108**) is obtained.[20] This drug is interestingly rapidly converted to **102** soon after ingestion and is in effect simply a prodrug for the latter.

Drugs that inhibit the conversion of angiotensin 1 to the vasoconstricting angiotensin 2, the so-called angiotensin converting enzyme (ACE) inhibitors, block the action of angiotensin converting enzyme, one of a series of zinc metalloproteases. A closely related enzyme causes the degradation of the vasodilating atrial natriuretic peptide. A compound that blocks both metalloproteases should in principle lower vascular resistance and thus blood pressure by complementary mechanisms. A drug that combines those actions, based on a fused two-ring heterocyclic nucleus, omapatrilat, is described in Chapter 10. A related compound that incorporates a single azepinone ring shows much the same activity. The synthesis begins by Swern oxidation of the terminal alcohol in the heptanoic ester **109**. Reaction of the product **110** with trimethylaluminum proceeds exclusively at the aldehyde to afford the methyl addition product (**111**). A second Swern oxidation, flowed this time by methyl titanium chloride, adds a second methyl group to afford the *gem*-dimethyl derivative (**112**). Construction of the azepinone ring begins by replacement of the tertiary carbinol in **112** with an azide group by reaction with trimethylsilyl azide and boron trifluoride. Hydrogenation of the product (**113**) reduces the azide to a primary amine and at the same time cleaves the benzyl ester to the corresponding acid (**114**). Treatment of this intermediate with a diimide leads to formation of an amide, and thus the desired azepinone ring (**115**). The phthalimido function, which has remained intact through the preceding sequence, is now cleaved in the usual way by reaction with hydrazine. The newly freed amine is again protected, this time as it triphenylmethyl derivative. The anion on the amide nitrogen from treatment of **116** with lithium hexamethyl disilazane is then alkylated with ethylbromoacetate; exposure to trifluoracetic acid (TFA) then cleaves the protecting group on the other nitrogen to afford **117**. The primary amino group is acylated with (S)acetylthiocinnamic acid (**118**). Saponification cleaves both the acetyl protection group on sulfur and the side-chain ethyl ester to afford **gemopatrilat (119).**[21]

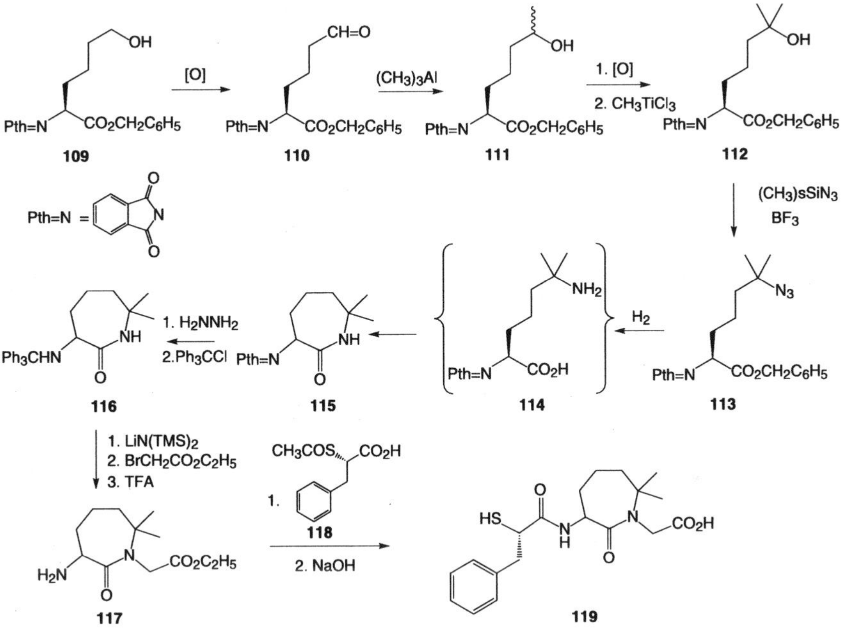

REFERENCES

1. D.P. Getman et al., *J. Med. Chem.* **36**, 288 (1993).
2. R.D. Tung, M.A. Murcko, G.R. Bhisetti, U.S. Patent 5,558,397 (1996). The scheme shown here is partly based on that used to prepare darunavir and phosamprenavir due to difficulty in deciphering the patent.
3. A.K. Ghosh, Y. Chen, *Tetrahedron Lett.* **36**, 505 (1995).
4. D.L.N.G. Surleraux et al., *J. Med. Chem.* **48**, 1813 (2005).
5. L.A. Sobrera, L. Martin, J. Castaner, *Drugs Future,* **23**, 22 (2001).

6. G. Bold et al., *J. Med. Chem.* **41**, 3387 (1998).

7. For a scheme for this intermediate see D. Lednicer, "The Organic Chemistry of Drug Synthesis", Vol. 6, John Wiley & Sons, Inc., NY 1999, pp. 12,13.

8. E.J. Stoner et al., *Org. Process Res. Dev.* **4**, 264 (2000).

9. P.S. Dragovich et al., *J. Med. Chem.* **42**, 1213 (1999).

10. A. Zask et al., *J. Med. Chem.* **47**, 4774 (2004).

11. L.M. Kauvar, M.H. Lyttle, A. Satyam, U.S. Patent 5,556,942 (1996).

12. A. Satyam, M.D. Hocker, K.A. Kane-Maguire, A.S. Morgan, H.O. Villar, M.H. Lyttle, *J. Med. Chem.* **39**, 1736 (1996).

13. J.A. McIntyre, J. Castaner, *Drugs Future* **29**, 992 (2004).

14. A. Sartorelli, K. Shyam, U.S. Patent 4,684,747 (1987).

15. A. Sartorelli, K. Shyam, P.G. Penketh, U.S. Patent 5,637,619 (1997).

16. J.P. Durlach, U.S. Patent 4,355,043 (1982).

17. S. Toyoshima, Y. Seto, U.S. Patent, 4,816,484 (1989).

18. W. Grell, R. Hurnaus, G. Griss, R. Sauter, M. Reiffen, E. Rupprecht, U.S. Patent, 5,216,167 (1993).

19. *See* Ref. 7, pp. 15–18.

20. L.A. Sobrera, J. Castaner, *Drugs Future*, **27**, 201 (2002).

21. J.A. Robl et al., *J. Med. Chem.* **42**, 305 (1999).

CHAPTER 2

ALICYCLIC COMPOUNDS

The slimness of this chapter very aptly reflects the importance of aromatic and heterocyclic moieties as cores for therapeutic agents. This section includes several agents that depend on the presence of on a single alicyclic group for their activity. Though a few of the compounds included in this chapter do include a benzene ring, that group does not seem to play a major role in their biological activity. Sizeable chapters were devoted in the earlier volumes in this series to the discussion of drugs based on the steroid nucleus. This area, in common with the prostaglandins that open this chapter, has received relatively little attention in recent years. A handful of steroids thus round out this section.

1. MONOCYCLIC COMPOUNDS

A. Prostaglandins

The discovery of the prostaglandins in the mid-1960s led to an enormous amount of research in both industrial and academic laboratories.[1] Much of this work was arguably based on the mistaken premise that these hormone-like compounds would provide the basis for the design of major new classes of therapeutic agents. Analogy with the large number of important

The Organic Chemistry of Drug Synthesis, Volume 7. By Daniel Lednicer
Copyright © 2008 John Wiley & Sons, Inc.

drugs that had emerged from manipulation of the structure of the steroid hormones provided at least some of the impetus for research on the prostaglandins. The expectation of major classes of new drugs was to some extent dispelled by the findings in the late 1960s that prostaglandins and other products derived from arachidonic acid tended to mediate injurious responses, such as inflammation rather than hormonal effects. In spite of this, several compounds based on this structure have found uses in the clinic. These include, for example, misoprostol, a drug based on the beneficial effect of this class of compounds on the mucous lining of the stomach. This compound differs from the naturally occurring prostaglandin E (PGE) by the removal of the side chain hydroxyl to one carbon furthest from the cyclopentane and presence of an additional methyl group on that position. Other compounds based on this structure provide several ophthalmic products. These topical prostaglandins are used to lower intraocular pressure in glaucoma patients. They are believed to cause the outflow of fluid within the eye by binding to specific intraocular receptors. These compounds are classed as PGFs. Since both ring oxygens comprise alcohols.

Misoprotol

The first two agents, both used to lower intraocular pressure due to glaucoma, differ from natural prostaglandins by the presence of an aromatic ring at the end of the neutral side chain. Construction of the terminal group on **travoprost** (**9**) starts with the reaction of the anion from methyl dimethoxyphophonate with the aryloxy acid chloride (**1**). Displacement of halogen affords product **2**. Treatment of **2** with strong base leads to formation of the corresponding ylide. A key intermediate for the synthesis of many prostaglandins is the so-called Corey lactone. This compound, shown as its dibenzylcarboxylic ester (**3**), provides functionality for immediate attachment of the neutral side chain as well as, in latent form a junction point for the addition of the carboxylic acid bearing side chain. Reaction of this reactive species from **2** with **3** leads to the condensation product of the ylide with the aldehyde. The trans stereochemistry of the newly formed olefin follows from the normal course of this reaction. Reduction of the side-chain carbonyl group with zinc borohydride gives the alcohol **5** after separation of the isomers. Reaction with mild base results in hydrolysis of the diphenyl ester protecting group (**6**). The newly revealed alcohol, as well as that on the side chain, are then protected as their tetrahydropyranyl ethers by reaction with dihydropyran in the presence of acid (**7**). Controlled reduction of the lactone ring with diisobutylaluminum hydride (DIBAL-H) stops at the lactol stage (**8**). Condensation with the zwitterionic salt-free ylide, 4-triphenylphosphinobutyrate, results in condensation with the open aldehyde form of the lactol. The "salt free" conditions of this reaction account for the formation of the new olefin with cis stereochemistry. The acid is then converted to its isopropyl ester by reaction of the carboxylate salt with isopropyl iodide. There is thus obtained travaprost (**9**).[2]

The rather simpler ophthalmic prostaglandin **latanoprost** (**10**),[3] carries a simple benzene ring directly at the end of the side chain. The amide of this drug is said to be better tolerated because of the absence of an acidic carboxyl group. Reaction of **10** with 1,8-diazobicyclo[5.4.0]undec-7-ene

(DBU) flowed by methyl iodide leads to the corresponding methyl ester **11**. Heating **11** with ethylamine leads to ester interchange and formation of the ethyl amide **bimatoprost (12)**.[4]

A PGE related compound, **lubiprostone (23)** displays a quite different spectrum of activity. This compound has been recently approved for treatment of chronic constipation and is being investigated for its effect on constipation-predominant irritable bowel syndrome. It has been ascertained that the drug interacts with specific ion channels in the GI tract causing increased fluid output into the lumen. Starting material for the synthesis (**13**) comprises yet another variant on the Corey lactone. Condensation of this aldehyde with the ylide from the difluorinated phosphonate **14** leads to the addition product **15**. The double bond in thef olefin has the expected trans geometry though the next step, hydrogenation, makes this point moot. Sodium borohydride then reduces the side-chain ketone function to give **17** as a mixture of isomers. The lactone is then reduced to the key lactol in the usual fashion by means of diisobutyl aluminum hydride. This compound is then condensed with the ylide obtained from reaction of the zwitterion 4-triphenylphosphoniumbutyrate to give the chain-extended olefin (**19**). The carboxylic acid in this intermediate is next protected as the benzyl ester by alkylation of its salt with benzyl chloride. Oxidation of the ring alcohol by means of chromium trioxide followed by exposure to mild acid to remove the tetrahydropyranyl group establishes the keto–alcohol PGE-like function in the five-membered ring (**21**). Catalytic hydrogenation of this last intermediate at the same time reduces the remaining double bond and removes the benzyl protecting group on the acid to give the open-chain version (**22**) of the product. The electron-withdrawing power of the fluorine atoms adjacent to the side-chain ketone cause the carbonyl carbon to become a reasonable electrophile. The electron-rich oxygen on the ring alcohol thus adds to this to give a cyclic hemiacetal. This form (**23**), greatly predominates in product **23**.[5]

B. Antiviral Agents

The enzyme neuraminidase plays a key role in the replication of influenza viruses. Newly formed virions form blister-like buds on the inside of the host cell membrane. This proteolytic enzyme, known as sialidase, facilitates scission of the base of the bud and thus release of the new virion. Research on structurally modified neuraminidase congeners culminated in the development of a derivative that blocked the action of the enzyme. This complex dihydropyran, zanamivir (**24**), was found to be effective against influenza viruses and most notably avian influenza A (H5N1), better known as bird flu. The lengthy complex route to that drug encouraged the search for alternative structures that would fit the same site. Two more easily accessible agents based on carbocyclic rings that have the same activity as the natural product based drugs have been identified to date.

There has been much recent anxious speculation that the bird flu virus will mutate to a form that will spread from person to person. A worldwide influenza epidemic, at worst comparable to that which followed the Great War, could well be the net result of such an event. The efficacy of the first carbocyclic neuraminidase inhibitor, **oseltamivir** (**34**), known familiarly by its trade name Tamiflu, has focused considerable attention on this drug. A significant amount of work has been devoted to its preparation in the expectation of the very large quantities that would be required in the event that the pandemic materializes. The choice of starting material is constrained by the fact that virtually every carbon atom on the six-membered ring in the molecule bears a chirally defined substituent. All the early syntheses start with either shikimic (25) or gallic acid from plant sources. A typical approach involves construction of the epoxide 27 as an early step. Shikimic acid is first converted to its ethyl ester. The two syn disposed adjacent hydroxyl groups are then tied up as the acetonide (26) by reaction with acetone. The remaining free alcohol is then converted to its mesylate by means of methanesulfonyl chloride. Hydrolysis of the mesylate followed by reaction with base leads to internal displacement of the mesylate and formation of the key epoxide (27). The free hydroxyl group is then protected as its methoxymethylene ether (MEM) (28). Reaction with the nucleophile (sodium azide) then serves to open the oxirane to give the azide (29). The next series of steps in essence establishes the presence of adjacent trans disposed amino groups. The free hydroxyl is thus first converted to a mesylate; catalytic hydrogenation then reduces the azide to a primary amine (30). This displaces the adjacent mesylate to form an aziridine (31). Treatment of this intermediate or a derivative with acetic anhydride then affords the activated intermediate (32).

Sodium azide again opens a strained reactive three-membered ring. There is thus obtained the acetamido azide (**33**). Repeat of the sequence, mesylate formation followed by catalytic hydrogenation leads to formation of a new aziridine (**33**), in which the ring has moved over by one carbon atom. Reaction of this last intermediate with 3-pentanol in the presence of base leads to opening the aziridine ring to give the corresponding ether. The azide in this last intermediate is again reduced with hydrogen. Finally (**34**) is obtained.[6–8]

The importance of this drug to meet a potential worldwide pandemic has attracted the attention of academic chemists. This finding has resulted in the development of a relatively short but elegant synthesis. The approach is notable in that it does not involve difficult to obtain natural product starting material and completely obviates the use of potentially explosive azide intermediates. The first step involves building the carbocyclic ring equipped with a chiral carbon atom that will determine the stereochemistry of the many remaining ring substituents. Thus, $2 + 4$ cycloaddition of acrylate **5** to cyclobutadiene in the presence of the pyrrolidine derivative (**36**) as a chiral catalyst affords the ester (**37**) as a virtually pure optical isomer. The ester group is then converted to an amide (**38**) by simple interchange with ammonia. Reaction with iodine under special conditions leads to the nitrogen counterpart of an internal iodo-lactonization reaction and formation of the bridged iodolactam (**39**). The amide nitrogen is then protected as its *tert*-butoxycarbonyl derivative. Dehydroiodination with DBU introduces a double bond giving **41**. Free radical bromination of that intermediate with *N*-bromosuccinimide (NBS) proceed with the shift of the olefin to give the allylic bromide **42**. Dehydrobromination with cesium oxide introduces an additional double bond in the ring. The presence of ethanol leads the lactam ring to open at the same time. (For ease in visualization, the product diene (**43**) is drawn both as produced and in the orientation that matches that in the scheme above.) The second nitrogen required for the product is introduced by an unusual novel reaction. The product (**44**) obtained from treatment of diene **43** with acetonitrile and NBS in the presence of stannic bromide can be rationalized by positing initial addition of bromide to the olefin to form a cyclic bromonium ion; addition of the unshared pair of electrons on the nitrile nitrogen would account for the connectivity. Hydrolysis of the initial imine-like intermediate would account for the observed product **44**. Treatment with base leads to internal displacement and formation of the aziridine ring in **45**. Reaction as above with 3-pentanol followed by removal of the *t*-BOC group affords oseltamivir (**34**).[9]

Neuraminidase blocking activity is interestingly retained when the central ring is contracted by one carbon atom. Note that the cyclopentane ring in the antiviral agent **peramivir** (**53**) carries much the same substituents as its cyclohexane-based counterpart. The presence of the guanidine substituent, however, traces back to the tertrahydropyran zanamivir (**24**). The relatively concise synthesis of peramivir starts by methanolysis of the commercially available bicyclic lactam **46**. Reaction of the thus-obtained amino-ester with *t*-BOC anhydride leads to the *N*-protected intermediate (**47**). The key reaction involves addition of both a functionalized carbon substituent and a hydroxyl group in a single step. Reaction of the nitroalkane (**48**) with phenyl isocyanate leads to the formation of a nitrone. That very reactive species then undergoes 2 + 3 cycloaddition to the double bond in the cyclopentene (**47**). The isoxazolidine (**49**) is the predominant isomer from that reaction. Catalytic hydrogenation then cleaves the scissile nitrogen-to-oxygen bond leading to ring opening and formation of the corresponding aminoalcohol. This compound is converted to acetamide (**50**) with acetic anhydride. The ring amino group is next revealed by removal of the *t*-BOC group by means of acid to yield **51**. An exchange reaction of that primary amine with pyrazole carboxamidine (**52**) then introduces the guanidine group. Thus the antiviral compound **53** is obtained.[10]

HN
O
46

1. CH₃OH
2. (*t*-BuOCO)₂O

t-BOCNH   CO₂CH₃
47

NO₂   →   N—OH
48

C₆H₅N=C=O

t-BOCNH   CO₂CH₃
N
O
49

1. H₂
2. Ac₂O

*t*BOCNH   CO₂CH₃
OH
NHAc
50

H⁺

H₂N   CO₂CH₃
OH
NHAc
51

NH
H₂N   NH
CO₂CH₃
OH
NHAc
53

N
N
H₂N   NH
52

C. Miscellaneous Monocyclic Compounds

The venerable drug theophylline has been used to treat asthmatic attacks for many years. Research on the mechanism of action of this purine led to the discovery that it acted by blocking the action of the enzyme phosphodiesterase (PDE). The clinical utility of this compound pointed to the potential of PDE inhibitors as a source for new drugs. Modern methodology has led to the subdivision of PDE receptors into a number of subtypes, each of which seems to be involved in the regulation of a discrete organ system. It is of interest that a newly developed PDE inhibitor, in this case specific for PDE 4, has shown clinical activity in alleviating the symptoms of chronic obstructive pulmonary disease. This finding of course involves the same organ that led to the discovery of the field many decades ago. The first sequence in the synthesis of this new PDE inhibitor comprises homologation of the benzaldehyde (**54**) to phenylacetonitrile (**56**). Reduction of the aldehyde in the presence of lithium bromide gives the corresponding benzyl bromide; displacement of halogen by reaction with potassium cyanide gives the substituted acetonitrile (**56**). Condensation of that intermediate with ethyl acrylate in the presence of base leads to Michael addition of two molecules of the acrylate and formation of the diester (**57**). Treatment of this last intermediate with strong base leads to internal Claisen condensation with consequent formation of the carberthoxycyclohexanone derivative **58**. Heating the ketoester with acid initially leads to hydrolysis of the ester to an acid. This decarboxylates under reaction conditions to give the cyclohexanone (**59**). Condensation of the ring carbonyl group in this intermediate with the anion from 1,3-dithiane leads to **60**, in what in effect comprises addition to the ring of a carbon atom at the acid oxidation stage. Oxidation of the sulfur atoms in this intermediate with mercuric chloride in the presence of methanol converts that carbon to the methyl ester. The product in which the nitrile and carboxyl are cis to each other predominate in an 11 : 1 ratio over the trans isomer, probably reflecting thermodynamic factors. Saponifaction of the ester group completes the synthesis of **cilomilast (61)**.[11]

The vitamin A related compound, all trans retinoic acid, is a ligand for cells involved in epithelial cell differentiation. It has found clinical use under the name tretinoin (**62**), for treatment of skin diseases, such as acne, and for off-label application as a skin rejuvenation agent. The recently discovered closely related 9-cis isomer in addition binds to a different set of receptors involved in skin cell growth. This new compound has been found to control proliferation of some cancer cells. The drug is thus indicated for topical use in controlling the spread of Kaposi sarcoma lesions. Reduction of the ester group in compound **63**, which incorporates the requisite future 9-cis linkage with lithium aluminum hydride, leads to the corresponding alcohol. This compound is then oxidized to aldehyde **64** by means of manganese dioxide. Condensation of the carbonyl group with the ylide derived by treatment of the complex phosphonate (**65**) adds the rest of the carbon skeleton (**66**). Saponification of the ester then gives the corresponding acid, **alitretinoin (67)**.[12]

2. POLYCYCLIC COMPOUNDS: STEROIDS

The first five volumes in this series each feature a separate chapter that bears the title Steroids. The steady diminution of research on this structural class is illustrated by the regularly decreasing extent of those chapters. By the time Volume 6 appeared, the discussion of steroid-based drugs takes up but a section in the chapter on Carbocyclic Compounds. The relatively small amount of research devoted to this area is reflected in the present volume as well. The compounds that follow are organized by structural class as they each display quite different biological activities from one another.

A. 19-Nor Steroids

Virtually all known antiestrogens comprise non-steroidal compounds based more or less closely on triphenyl ethylene (see ospemifene, Chapter 3). Tamoxifen ranks as the success story in this class of drugs having found widespread use in the treatment of women suffering from estrogen receptor positive breast cancer. It is thus notable that a derivative of estradiol itself, which carries a very unusual substituent, also acts as an estrogen antagonist. Unlike its non-steroid forerunners, all of which show some measure of agonist activity, this agent is devoid of any estrogenic activity. Conjugate 1,4 addition of the Grignard reagent from long-chain bromide **69** to the dienone **68** affords the 7-alkylated product **70** as a mixture of epimers at the new linkage. The 7-α isomer is separated from the mixture before proceeding; the hydroxyl group at the end of the long chain is then unmasked by hydrolytic removal of the silyl protecting group. Reaction with cupric bromide serves to aromatize ring A; saponification with mild base preferentially removes the less sterically hindered acetate at the end of the long chain. Acylation by means of benzoyl chloride converts the phenol to the corresponding ester **72**.[13] The synthetic route from this point on is speculative as details are not readily accessible. Reaction of this last literature intermediate (**72**) with methanesulfonyl chloride would then lead to mesylate **73**. This group could then be displaced by sulfur on the fluorinated mercaptan (**74**). Controlled oxidation of sulfur, for example, with periodate, would then lead to the corresponding sulfone. Saponification of the ester groups at the 3 and 17 positions would afford the estrogen antagonist **fluverestant, (75)**.[13] Note that this specific compound is but one from a sizeable group of similarly substituted estradiol derivatives that shows pure antagonist activity.

Synthesis of 19-norandrostanes, which carried a functionalized benzene ring at the 11 position, led to the surprising discovery of compounds that antagonized the action of progestins and glucocorticoids. One of those compounds, mifepristone (**76**), more familiarly known as RU-486, was quickly enveloped in controversy, as a result of its use for terminating pregnancy by nonsurgical means. A more recent example, **asoprinosil (84)** is a more selective progesterone antagonist with reduced binding to glucocorticoid receptors. Treatment of the diene **77**, a total synthesis derived 19-nor gonane, with hydrogen peroxide, results in selective reaction at the 5,10 double bond to give the epoxide **78**. The proximity of the acetal oxygens at the 3 position accounts for the regiochemistry of the product; attack from the more open α-side accounts for the stereochemistry. Condensation of **78** with the Grignard reagent from the methyl acetal of *p*-bromobenzaldehyde in the presence of cuprous salts leads to conjugate addition to the 11 position with concomitant shift of the double bond to the 5,9 position (**79**). Construction of the side chain at the 17 position starts with the addition of the ylide from the tri-methylsulfonium iodide to the carbonyl group give the oxirane **80**. Reaction of the product with sodium methoxide opens the epoxide ring to give the ether–alcohol **81**. Alkylation of the hydroxyl group at 17 with methyl iodide and base affords the bis(methyl ether) **82**. Exposure of that intermediate to acid leads to hydrolysis of the acetal protecting groups on the ketone at 3 as well as the aromatic aldehyde; the tertiary alcohol on the AB ring junction dehydrates under those conditions, restoring the double bond (**83**). Reaction of this last intermediate with hydroxylamine leads to formation of the aldoxime in a 95 : 5 (*E*)/(*Z*) ratio.

The purified isomer (**84**)[14] is currently used in the clinic as a treatment for endometriosis, as well as other conditions related to excess progesterone stimulation.

The total synthesis based 17-ethyl norgonane, levonorgestrel (**85**), has been in use for many years. The corresponding oxime has recently been introduced as the progestational component of an oral contraceptive. In the absence of a specific reference, it may be speculated that the compound **norelgestimate** (**86**), is prepared by simply reactioning **85** with hydroxylamine in analogy to the preparation of the corresponding 17 acetoxy derivative.[15]

85   86

B. Corticosteroid Related Compounds

The soybean sterols that comprise one of the principal sources for progestins and corticoids consist almost entirely of a mixture of stigmasterol and sitosterol. The lack of a double bond in the side chain renders the latter useless as a starting material since that function is key to the Upjohn process for removing the long terpinoid side chain present in these sterols. As a result, tonnage quantities of this steroid thus accumulated in a Kalamazoo storage lot over the years. A microorganism was finally found in the late 1970s that would feed on this rich source of carbon, metabolizing the fatty side chain at the 17 position and at the same time introducing an oxygen function at position 9 that could be used later to introduce the crucial double bond in ring C. Considerable work was then devoted to developing methods for building up the highly functionalized two-carbon side chains found in corticoids at the now bare 17 position found in these biodegradation products. One of the intermediates from one of those schemes has interestingly recently becomes a drug in its own right. Wet macular degeneration, one variant of the disease that leads to loss of vision with age, is marked by excessive growth of new blood vessels in the retina. This process, neovascularization, in effect destroys photoreceptor areas of the retina. The drug, **anecortave** (**95**), which needs to be administered locally within the eyeball, inhibits the growth of those new blood vessels thus saving still-intact areas the retina. The sequence for preparation of the agent begins with the conversion of the ketone at 3 to its enol ether **88**, for example, with methyl orthoformate. Addition of cyanide to the carbonyl at 17 initially leads to a mixture of epimeric cyanohydrins. Conditions were developed for converting this to the desired isomer **89**, by crystallization under equilibrating conditions. The 17β cyano isomer is apparently less soluble that its epimers. The hydroxyl group at 17 is then protected as its trimethylsylyl ether (**90**). Reduction by means of diisobutylaluminum hydride followed by protic workup converts the cyano group to an aldehyde. Condensation of this intermediate with the anion from methylene bromide probably leads initially to adduct **92**. Excess base, lithium diisopropylamide (LDA), is thought to remove one of the bromine atoms. The resulting intermediate then rearranges to give the observed bromoketone. Removal of the protecting groups leads to diketone **94**, which lacks only oxygen at positon 21. Displacement of bromine atom at that position with sodium acetate completes construction of the corticosteroid side chain at position 17.[16] Thus **96** is obtained.

Topically applied corticosteroids have proven very effective in treating asthma. The amount of drug from the inhaled dose that reaches the small airways has a critical effect on relieving asthmatic episodes. The acetal at positions 16,17 of the diol (**97**) with acetone,[17] **desonide**, has long been indicated for treating asthma. It has recently been found that the acetal containing cylcohexane carboxaldehyde penetrates further into the small airways. The first step in this brief sequence comprises reaction of the syn diol (**96**) with isobutyric anhydride to afford the triply esterified intermediate **97**. Reaction of this compound with cylcohexane carboxaldehyde in aprotic solvent in the presence of acid leads to formation of the acetal by a transesterification-like sequence. The presence of the very bulky alkyl groups on the esters favors formation of that isomer of the acetal in which the cyclohexane is oriented away from the face of the steroid. The product (**98**) thus predominantly comprises the desired isomer. Saponication of the remaining ester at 17 followed by separation from the small amount of epimeric acetal then affords **cicledesonide (99)**.[18]

Replacement of carbon 4 in androstane by nitrogen leads to androgen antagonists. These drugs have proven useful for treating a condition marked by excess androgen or sensitivity there to such a benign prostatic hypertrophy and even hair loss. Earlier examples are covered in Volumes 5 and 6 of this series. Synthesis of the most recent example starts by formation of the amide between steroid carboxylic acid **100** and 1,4-bis(trifluoromethylaniline) (**101**) via the acid chloride. Ring A is then opened to the corresponding keto acid (**102**) by oxidation of the double bond in ring A by means of permanganate. Reaction of that compound with ammonia at elevated temperature can be visualized as proceeding through an amide–enamine, such as **103**. Amide interchange closes the ring to form the enamide **104**. Catalytic hydrogenation followed by reaction with DDQ completes the synthesis of **dutasteride (105)**.[19]

Vitamin D actually consists of a set of closely related compounds whose structures are based on the steroid nucleus with an opened ring B. These compounds control various metabolic processes from bone deposition to skin growth. One of the vitamins (D3), calcitrol, has been used to treat psoriasis and acne. A recent semisynthetic Vitamin D congener has shown an improved therapeutic index over the natural product. The synthetic sequence to this analogue hinges on selective scission of the isolated double bond in the side chain. The key step thus involves inactivation of the conjugated diene centered on ring A. This is accomplished by formation of a Diels–Alder-like adduct between the starting material **106**, in which the hydroxyl groups are protected as the diisopropylsilyl ether and sulfur dioxide.[20] Ozonolysis of the adduct **107** followed by work-up of the ozonide, gives the chain-shortened aldehyde. Heating the product restores the diene by reversing the original cylcoaddition reaction (**108**). The carbonyl group is then reduced to the alcohol by means of borohydride and the resulting alcohol converted to its mesylate **109**. The chain is next homologated by first displacing the mesylate with the anion form diethyl malonate. Saponification of the ester groups followed by heating the product in acid cause the malonic acid to lose carbon dioxide. Thus, the chain extended acid **110** is obtained. The carboxyl group is next converted to the activated imidazolide by reaction with carbonyl diimidazole. Condensation of this intermediate with pyrrolidine gives the corresponding amide. Removal of the silyl protection groups with fluoride leads to **ecalcidine (111)**.[21]

3. POLYCYCLIC COMPOUNDS

A sequiterpinoid fulvene from the Jack O'Lantern mushroom was found some years ago to show very promising antitumor activity in a number of *in vitro* assays. This compound, Iludin S (**112**) was, however, found to be too toxic in follow-up *in vivo* tests to be considered for further development. The semisynthetic analogue **irofulven (114)** is far better tolerated and has been taken to the clinic where it demonstrated activity against several tumors. Though a total synthesis has been published,[22] the compound is most efficiently prepared by semisynthesis from Iludin S itself. Treatment of the natural product with dilute acid leads to loss of the elements of water and formaldehyde in what may be considered a reverse hydroformylation reaction to yield intermediate **113**. Reaction of that product with formaldehyde in dilute acid in essence reverses the last reaction, though with a different regiochemistry. Formaldehyde thus adds to the last open position of the extended eneone system. Thus, **114** is obtained.[23]

Pain receptors become sensitized during inflammation or from a related stimuli, and consequently often release glutamate. That amino acid then acts on NMDA receptors on neurons, and in the process further sensitizes those to pain stimuli. This sequence then causes the pain to become chronic. Specific NMDA antagonists would thus be expected to relieve chronic pain by interrupting that chain. Consequently, such compounds would offer a potentially nonaddictive alternative to the opiates currently used to treat chronic pain. Reductive amination of the acetaldehyde derivative **116** with monocarbobenzyloxy ethylenediamine (**116**) leads to the now-disubstituted ethylenediamine **116**. Reaction of this amine with the commercially available cyclo-butenedione derivative **117** leads to replacement of one of the ethoxy groups in **117** by the free amino group in **116** by what is probably an addition–elimination sequence to afford the coupled product **118**.

Reduction of this intermediate by hydrogen transfer from 1,4-cyclohexadiene in the presence of platinum leads to loss of the carbobenzoxy group and formation of the transient primary amine **119**. The terminal primary amino group in that product then participates in a second addition–elimination sequence to form an eight-membered ring (**120**). Treatment of this intermediate with trimethylsilyl bromide then cleaves the ethyl ethers on phosphorus to give the free phosphonic acid and thus **perzinfotel** (**121**).[24]

REFERENCES

1. See, for example, *Chemistry of the Prostaglandins and Leukotrienes*, J.E. Pike, D.R. Morton, Eds., Raven Press, NY, 1985.

2. J. Castaner, L.A. Sobrera, *Drugs Future* **25**, 41 (2000).

3. See D. Lednicer, *The Organic Chemistry of Drug Synthesis*, Vol. 6, John Wiley & Sons, Inc., NY 1999, p. 98.

4. D.A. Woodward, S.W. Andrews, R.M. Burk, M.E. Garst, U.S. Patent 5,352,708 (1994).

5. L.A. Sobrera, J. Castaner, *Drugs Future* **29**, 336 (2004).

6. C.U. Kim et al., *J. Am. Chem. Soc.* **119**, 681 (1997).

7. C.U. Kim et al., *J. Med. Chem.* **41**, 2451 (1998).

8. N. Bischofberger, C.U. Kim, W. Lew, H. Liu M.A. Williams, U.S. Patent 5,763,483 (1998).

9. Y.-Y. Yeung, S. Hong, E.J. Corey, *J. Am. Chem. Soc.* **128**, 6310 (2006).

10. Y.S. Babu et al., *J. Med. Chem.* **43**, 3482 (2000).

11. S.B. Christensen et al., *J. Med. Chem.* **41**, 821 (1998).

12. M.F. Boehm, M.R. McClurg, C. Pathirana, D. Mangelsdorf, S.K. White, J. Herbert, D. Winn, M.E. Goldman, R.A. Hayman, *J. Med. Chem.* **37**, 408 (1994).

13. J. Bowler, T.J. Lilley, J.D. Pittman, A.E. Wakelin, *Steroids* **54**, 71 (1989).

14. G. Schubert, W. Eiger, G. Kaufmann, B. Schneider, G. Reddersen, K. Chwalisz, *Semin. Reprod. Med.* **23**, 58 (2005).

15. A.P. Shroff, U.S. Patent 4,027,019 (1977).

16. J.G. Reid, T. Debiak-Krook, *Tetrahedron Lett.* **31**, 3669 (1990).

17. S. Bernstein, R. Littell, *J. Am. Chem. Soc.* **81**, 4573 (1959).

18. J. Calatayud, J.R. Conde, M. Lunn, U.S. Patent 5,482,934 (1996).

19. K.W. Batchelor, S.W. Frye, G.F. Dorsey, R.A. Mook, U.S. Patent 5,565,467 (1996).

20. R. Hesse, U.S. Patent 4,772,433 (1988).
21. R.H. Hesse, G.S. Reddy, S.K.S. Setty, U.S. Patent 5,494,905 (1996).
22. T.C. McMorris, Y. Hu, M. Kellner, *Chem. Commun.* 315 (1997).
23. T.C. McMorris, M.D. Staake, M.J. Kellner, *J. Org. Chem.* **69**, 619 (2004).
24. W.A. Kinney et al., *J. Med. Chem.* **41**, 236 (1998).

CHAPTER 3

MONOCYCLIC AROMATIC COMPOUNDS

Free-standing benzene rings have provided the core for a very large number of biologically active compounds. This ubiquitous unit provides the rigid flat skeleton on which to attach functional groups and also manifests the electron density required for recognition at various receptor sites. The ubiquitous occurrence of aromatic rings in endogenous effector molecules as, for example, the chatecholamines, is a reflection of this importance. The 30 odd drugs described in Chapter 3 represent a wide variety of structural types; their biological activities are equally diverse. Consequently, it is often not readily apparent just which parts of the structures form part of the pharmacophore. Thus, compounds are included in this chapter largely on the basis of structure. The grouping in the discussions that follow are admittedly somewhat arbitrary.

1. ARYLCARBONYL DERIVATIVES

A relatively simple benzamide inhibitor of phosphodiesterase 4 has proven useful for treating congestive obstructive pulmonary disease (COPD). The synthesis of this compound begins by alkylation of the free phenol on aldehyde **1** with chlorodifluoromethane in the presence of base. Oxidation of

The Organic Chemistry of Drug Synthesis, Volume 7. By Daniel Lednicer
Copyright © 2008 John Wiley & Sons, Inc.

the carbonyl group in **2** with hypochlorite followed by reaction with thionyl chloride leads to the acid chloride (**3**). Condensation of the last intermediate (**3**) with the substituted aminopyridine (**4**) affords the benzamide **roflumilast** (**5**).[1]

The discovery of the "statin" mevalonic acid synthesis inhibitors focused new attention on control of blood lipid levels as a measure to stave off heart disease. A number of compounds have been found that treat elevated lipid levels by other diverse mechanisms. The phosphonic acid derivative **ibrolipim** (**9**) is believed to lower those levels by accelerating fatty acid oxidation. The phosphorus-containing starting material **7** can in principle be obtained by the Arbuzov reaction of a protected from of *p*-bromomethylbenzoic acid (**6**) with triethyl phosphate. Removal of the protecting group and conversion of the acid to an acyl chloride then affords **7**. Condensation of this intermediate with substituted aniline **8** leads to the hypolipidemic agent (**9**).[2]

Red blood cells in patients afflicted with sickle-cell anemia show a reduced capacity for holding oxygen. The adventitious discovery that the lipid lowering agent clofibrate (**10**) possessed some antisickling activity led to further investigation of related compounds. It was found that this unexpected activity of these agents resulted from their bringing about an increased oxygen-carrying capacity of normal blood cells. One of the compounds from this work, **efaproxiral (15)**, is now used to deliver oxygen to hypoxic tumor tissues to improve the efficacy of radiation therapy. Oxygen, whose presence is crucial to the generation of cell-killing radicals, is often in short supply in solid tumors. Reaction of the arylacetic acid (**11**) with thionyl chloride leads to the corresponding acyl chloride. Condensation of that intermediate with 3,5-dimethylaniline leads to the anilide **14**. Treatment of that product with acetone and chloroform in the presence of strong aqueous base adds the dimethyl acyl function to the phenol group. Thus, **15** is obtained.[3] This unusual reaction can be visualized by assuming intitial formation of a hemiacetal between the phenol and acetone (a); displacement of the hydroxyl by the anion from chloroform would lead to the intermediate (b); simple hydrolysis would then convert that group to a carboxylic acid.

Over the past few years, it has been established that several apparently quite unrelated drug classes owe their activity to effects on a shared biochemical system. The blood lipid lowering effect of the fibrates, such as, **10**, and the hypoglycemic action of the recently introduced hypoglycemic thiazolidine-diones both trace back to action on subtypes of the peroxisome proliferator-activated receptors (PPAR), which regulates lipid and glucose metabolism. Research targeted at PPAR has led to several novel hypoglycemic agents, which are unrelated structurally to drugs previous used to treat diabetes.

The synthesis of the first of these starts with the formation of the enamide (**18**) from tyrosine (**16**) and 2-benzoylcyclohexanone (**17**). Treatment of that product with palladium on charcoal leads to dehydrogenation of the ene-amide ring with consequent aromatization. Condensation of the terminal hydroxyl group on the side chain in the substituted oxazole (**21**) with the phenol function on **20** in the presence of triphenylphosphine and diethyl azo-dicorboxylate (DEAD) leads to formation of an ether bond. This reaction affords the hypoglycemic agent **farglitazar (22)**.[4]

An aryl carbamate replaces the tyrosine moiety in a related analogue. The preparation of this compound first involves activation of the side-chain oxygen in the same oxazole used above (**21**) by conversion to its mesylate (**21a**) by means of methanesulfonyl chloride. This intermediate is used to alkylate the phenol group on *p*-hydroxybenzaldehyde (**23**). Condensation of the aldehyde group in **24** with glycine methyl ester leads to the corresponding imine. Reduction of that function with boro-hydride yields the intermediate **25**. Acylation of the amino group in **25** compound with *p*-anisyl chloroformate (**26**) yields **muraglitazar (27)**.[5]

A very simple triketone has proven useful for treating the rare genetic disease tyrosinemia. The drug alternately known as **nitisinone (30)** or **orfadin** actually bears a very close relation to a pesticide. The analogue in which methylsulfonyl replaces the trifluoromethyl group, mesotrione, is an important corn herbicide. Acylation of cyclohexan-1,3-dione, shown as an enol (**28**) with acid chloride (**29**), leads in a single step to **30**.[6]

28   **29**   **30**

2. BIPHENYLS

Excision of malignant tumors comprises first line treatment for cancer of solid tissues. This procedure not infrequently misses small fragments of the tumor that may have broken off before surgery from the principal site of the disease. These fragments, metasteses, often proliferate at quite remote locations where they cause much of the pathology of cancer. A series of proteolytic enzymes present in tumor cells, known as matrix metalloproteinases, help establish growth of these metasteses at the newly invaded sites; these proteases are also involved in the formation of new blood vessels that will nourish the invasive cell masses. Consequently, considerable research has been devoted to matrix metallo-proteinases as a target for anticancer drugs. Clinical results with these compounds have to date produced equivocal results.[7]

Construction of one biphenyl-based metaloproteinase inhibitor starts with the Friedel–Crafts acylation of 4-chlorobiphenyl (**31**) with itaconic anhydride (**32**). Attack proceeds on the less inactivated ring to give the acylated product (**33**). Michael addition of phenylmercaptan to the exomethylene group gives the proteinase inbitor **tanomastat (34)**.[8]

Research on the prostaglandins several decades ago revealed a second pathway in the arichidonic acids cascade. Products from this alternate route consisted of long-chain fatty acids instead of the cyclic prostanoids. Subsequently, it was ascertained that these straight-chain acids, called eicosanoids, reacted with endogenous sulfur-containing oligopeptides, such as, glutathione, to form a series of compounds called leukotrienes that elicited allergic reactions. These were found to be the same as the previously known slow-reacting substance of anaphylaxis (SRS-A). Drugs that counteract this factor would provide an alternate means to antihistamines for dealing with allergies. The majority of antagonists developed to date consist of long-chain compounds that terminate in an acidic function. This last comprises a tetrazole rather than a carboxylic acid in virtually all compounds reported to date. A very recent entry interestingly terminates in a carboxylic acid. The convergent synthesis starts with the alkylation of the enolate from resorcinol monomethylether (**35**) with propyl iodide. The use of aprotic conditions favors alkylation on carbon over oxygen to afford (**36**). The enolate from this product is then used to displace the halogen atom in *o*-fluorobenzonitrile to give the aromatic ether. The nitrile is then hydrolyzed to the corresponding acid with potassium hydroxide. Treatment of that product with methanol in the presence of acid converts the latter to its methyl ester. The *O*-methyl ether in the product is then cleaved with boron tribromide to afford the phenol (**39**). Synthesis of the second half of the compound begins with the reduction of the acetyl group in **40** to an ethyl group (**41**) with tirethylsilane. Bromination of this intermediate with NBS proceeds, as expected at the position adjacent to the benzyl ether to afford **42**. The side chain that will connect the two halves is then put in place by alkylating the free phenol group with 1-bromo-2-chloroethane (**43**). Reaction of the aromatic bromo function in this last intermediate with *p*-fluorophenylboronic acid (**44**) leads to formation of the biphenyl group and thus **45**. The Finkelstein reaction, with potassium iodide, leads to replacement of chlorine by the better leaving group, iodine in **46**. Alkylation of the enolate from the acid moiety **39** with iodide in **46** completes construction of the framework.

Saponification of the methyl ester followed by hydrogenolysis of the benzyl ether completes the synthesis of the leukotriene B$_4$ antagonist **etalocib** (**47**).[9]

The discovery, close to half a century ago that beta adrenergic receptors fell into two broad classes, led to major advances in drug therapy. Agonists that act on beta 2 receptors, for example, include the majority agents for treating asthma. The large number of beta blockers act as beta-1 selective adrenergic antagonist. The discovery of a third subset of binding sites, the beta-3 receptors has led to a compound that provides an alternate method to currently available anticholinergic agents for treating overactive bladders. There is some evidence too that beta-3 agonists may have some utility in treating Type II diabetes. Synthesis of the compound begins with construction of the biphenyl moiety. Thus, condensation of methyl *m*-bromobenzoate (**48**) with *m*-nitrophenylboronic acid (**49**) in the presence of palladium leads to the coupling product (**50**). The nitro group is then reduced to the corresponding amine (**51**). Alkylation of **51** with the *t*-BOC protected 2-bromoethylamine (**52**) leads to intermediate **53**. Treatment with acid removes the protecting group to give the primary amine (**54**). Condensation of this last product with *m*-chlorostyrene oxide leads to formation of **56**, a molecule that incorporates the aryl ethanolamine moiety present in the great majority of compounds that act on adrenergic receptors. Thus, **solabegron (56)** is obtained.[10]

50; R = O
51; R = H

3. COMPOUNDS RELATED TO ANILINE

The structures of many monocyclic aromatic compounds have little in common; the same is often the case for their biological activities. They are consequently grouped here on the basis of some shared structural element rather than biological activity. The three compounds that follow have in common a nitrogen atom attached to the benzene ring.

A wide variety of compounds were tested as lipid-lowering agents some five decades ago when association between heart disease and hyperlipdemia was established. One of the more effective agents discovered in the course of this research was the endogenous hormone thyroxin. Use of this compound was severely limited by its effect on metabolic function and cardiac activity. More detailed recent research has the thyroid receptor, like those for many other hormones, exists as a pair of subtypes that are unevenly distributed.[11] One of these, the β-receptor, is virtually absent in cardiac tissue. Research has thus started to identify compounds aimed at this receptor. The synthesis of a selective blocker for the β selective agent **axitirome (66)**, starts with the reaction of the dianisyl ioidonium salt (**57**) with the nitrophenol (**58**) in the presence of a cupric salt to give the product from displacement of iodine by the phenoxide to give the diaryl ether (**59**). Acylation of intermediate (**59**) with p-fluorobenzoyl chloride (**60**) in the presence of titanium tetrachloride yields the benzophenone (**61**). Reaction intermediate (**61**) with boron tribromide then serves to cleave the methyl ether (**62**). Catalytic hydrogenation serves to reduce the nitro group to afford aniline (**63**). This intermediate is then treated with ethyl oxalate (**64**) to afford the oxamic ester **65**. The carbonyl group is then reduced to the corresponding alcohol; saponification affords the hypolipidemic agent **66**.[12]

The few drugs currently available for treating hair loss act by very different mechanisms, Finasteride and structurally related steroids act to diminish the effect of testosterone, the hormone closely associated with male pattern hair loss. The mechanism of action of the older hair growth stimulant, minoxidil (**67**) on the other hand, is believed to involve the compound's vasodilatory activity. The more recent hair growth stimulant, **namindinil (74)**, also shows vasodilator action. It is of note that the nitrogen-rich functionality in the latter somewhat resembles a ring-opened version of **67**. The synthesis of the newer compound reflects the current trend to prepare drugs in chiral form. One branch in the convergent scheme thus involves preparation of the complex alkyl group as a single enantiomer. Thus reductive alkylation (R)-α-methylbenzylamine (**69**), with pinacolone (**68**), gives the secondary amine as a mixture of diastereomers. These are then separated by chromatography. Catalytic hydrogenation of the appropriate diastereomers leads to scission of the benzylic carbon-to-nitrogen bond to afford the (R)-amine (**71**) as a single isomer. Construction of the second part of the molecule starts by addition of sodium cyanamide to the isothiocyanate (**72**), itself, for example, available from reaction of the aniline with thiophosgene. The thiourea product (**73**) is then condensed with the chiral amine (**71**) to afford **74**.[13]

The activity against urinary incontinence of the adrenergic beta-3 receptor agonist solabegron (**56**) was noted above. An agent that acts on a subset of alpha receptors, specifically, alpha-1A/1L receptors, has also shown activity on the same clinical end point. The synthesis starts with Mitsonobu alkylation of the nitrophenol (**75**) with trityl protected imidazole carbinol (**76**) to yield the ether (**78**). The nitro group on the benzene ring is then reduced by any of several methods (**79**). The resulting aniline is then converted to the corresponding sulfonamide (**80**), by reaction with methanesulfonyl chloride. Hydrolysis with mild acid then removes the trityl protecting group to afford **dabuzalgron** (**81**).[14]

4. COMPOUNDS RELATED TO ARYLSULFONIC ACIDS

The title for this section aptly illustrates the almost arbitrary criteria used to group potential drugs in this chapter. The few entries in this section comprise an interesting contrast to a full sizeable chapter that was devoted to sulfonamide related drugs in Volume 1 of this series. Arylsulfonic acid derived moieties formed an essential part of the pharmacophore in virtually every one of the 40 odd antibacterial, diuretic, and antidiabetic agents described in that chapter. In the case at hand, by way of contrast, this functionality is no more than a convenient handle by which to coral a group of compounds with otherwise widely divergent structures, as well as biological activities.

A benzene ring that contains two sulfonic acid groups, as well as a nitrone, is currently being investigated as a treatment for stroke. The free radical scavenging properties of this compound, **disufenton (89)**, will, it is hoped, translate into neuroprotective action on brain tissue when administered during the first 6 h after a cerebral stroke. The synthesis of this compound starts with the preparation of the highly substituted hydroxylamine (**86**). Thus, reaction of benzaldehyde with *tert*-butylamine leads to imine **83**. Oxidation of intermediate **83** with *m*-perchlorobenzoic acid (MCPBA) gives the oxazirane (**84**); this rearranges to the corresponding nitrone (**85**) on heating. Hydrogen sulfide then serves to reduce that intermediate to the requisite hydroxylamine (**86**). In a somewhat unusual reaction, treatment of 2,4-dichlorobenzaldyde with sodium sulfite at elevated temperature leads to displacement of each of the halogens by sulfur to give the disulfonic acid (**88**) as its sodium salt. This S_N2-like reaction is probably facilitated by the lowered electron density at the 2 and 4 positions of the starting benzaldehyde (**87**). Reaction of the products (**88**) with the intermediate (**86**), leads to formation of the nitrone (**89**) by hydroxylamine interchange. Thus **89** is obtained.[15]

Elastase is an inflammatory protease associated with lung injury caused by infection or any of a number of other tissue insults. Inhibitors of this enzyme would thus be expected to aid in treatment of lung injuries. A compound that includes a sulfonamide linkage has shown activity against neutrophil elastase. Protection of the phenol in *p*-hydroxyphenylsulfonate (**90**) as its *tert*-butyl ester (**91**), comprises the first step in the construction of an elastase inhibitor. The sulfonate function is then activated as the sulfonyl chloride (**92**) by reaction with thionyl chloride. In the other arm of the convergent scheme, condensation of *o*-nitrobenzoyl chloride (**93**) with glycine benzyl ester (**94**) leads to the second half of the molecule. The nitro group in this intermediate is then reduced to an amine with iron in the presence of acid. Reaction of the newly revealed amino group with sulfonyl chloride in **92** yields the corresponding sulfonamide (**97**). Hydrogenolysis of the benzyl protecting group affords the free acid and thus **sivelestat** (**98**).[16]

The potential antimetastatic activity of inhibitors of metalloproteinases has generated considerable work in the area as noted in the discussion of tanomastat (**34**). The very different structure of the inhibitor **prinomastat** (**107**) illustrates the considerable structural freedom that seems to exist for inhibitors of these enzymes. Displacement of halogen in 4-chloropyridine (**99**) by phenoxide leads to the diaryl ether (**100**). Reaction of intermediate **100** with chlorosulfonic acid affords the sulfonyl chloride (**101**).

Construction of the thiomorpholine moiety starts by protection of the carboxylic acid in penicillamine (**102**) by reaction with hexyl dimethylsilane (Dmhs). Condensation of **103** with 1,2-dichloroethane leads to formation of the heterocyclic ring by sequential displacement of halogen by nitrogen and sulfur to yield **104**. Reaction of intermediate **104** with the benzophenone (**101**), leads to formation of the sulfonamide (**105**). Mild acid then serves to reveal the free carboxylic acid (**106**). This function is then converted to the acid chloride with oxalyl chloride. Treatment of this reactive intermediate with hydroxylamine leads to acylation on nitrogen to afford the proteinase inhibitor **107**.[17]

Endothelins comprise a group of vasoconstrictive peptides generated by the endothelium of blood vessels, as well as other tissues. Studies on inhibitors or frank antagonist suggest that such compounds will be of value in treating various cardiovascular diseases. The synthesis of an antagonist starts by forming a sulfonamide linkage. Thus, reaction of *o*-bromobenzenesulfonyl chloride (**108**) with aminoisoxazole (**109**) gives the sulfonamide (**110**). The somewhat acidic sulfonamide function is then tied up as its methoxymethylene (MOM) derivative. Lithium–halogen interchange of the bromine in **111** followed by reaction with trimethyl borate gives the borate ester (**112**). Mild acid leads to the corresponding boric acid derivative. Reaction of that compound with the substituted bromobenzene (**114**) in the presence of a palladium–triphenyl phosphine complex leads to aryl coupling and formation of the biphenyl (**115**). The MOM group is lost along the line, most likely during hydrolysis of the borate ester. Reductive amination of the aldehyde function in

115 with methylamine gives intermediate **116**. Acylation of the secondary amine with pivaloyl chloride affords the corresponding amide and thus the endothelin receptor antagonist **edonentan** (**117**).[18]

Cholesterol is not absorbed from the intestine as such, but needs to be esterified first. This process requires a special enzyme, ACAT (acyl-CoA : cholesterol acyltransferase). Inhibitors of this enzyme should provide a means of reducing serum cholesterol levels by limiting the amount absorbed from the diet. The inhibitor **avasamibe** (**125**) is included in this section though the sulfur-containing function, which comprises an unusual sulfonamide linked through oxygen rather than carbon. Chloromethylation of commercially available 1,3,5-triisopropylbenzene (**118**) with formaldehyde and hydrogen chloride affords the chloromethylated derivative (**119**). Displacement of the benzylic chlorine by cyanide gives the corresponding nitrile (**120**). The cyano group is then hydrolyzed with base to give the arylacetic acid (**121**). The other branch of the convergent synthesis starts with reaction of the hindered phenol (**122**) with isocyanosulfonyl chloride to give the O-sulfonated product (**123**). Hydrolysis then leads to O-sulfonamide (**124**). Acylation of intermediate **124** with the acid chloride obtained from acid **122** and thionyl chloride leads to the ACAT inhibitor (**125**).[19]

The pH of intracellular space is known to fall during ischemic episodes, heart surgery, and other events that compromise cardiac function. This encourages entry of sodium and calcium into cardiac cells. The resulting calcium overload may then compromise contractile function. The recently developed antiarrhythmic agent **cariporide (130)** has been found to act by specifically inhibiting the sodium–proton exchange that leads to further injury. Chlorosulfonation of *p*-isopropylbenzoic acid leads to the chlorosulfonyl derivative (**127**). Treatment of intermediate **127** with sodium sulfite in base serves to reduce the newly introduced function to the sulfenic acid (**128**). Alkylation by means of base and methyl bromide gives the corresponding aryl methyl sulfone (**129**). The carboxylic acid is then activated as the acid chloride by treatment with thionyl chloride. Reaction of this intermediate with guanidine gives the aroyl guanidine (**130**).[20]

5. DIARYLMETHANES

Prolonged treatment of cancer patients with chemotherapeutic agents not infrequently results in the development of drug resistance. The small number of malignant cells that survive exposure to the drug proliferate and become the dominant population. A far more serious case involves the development of cell lines that develop resistance to more than one class of drugs, a condition known as multidrug resistance. The structurally relatively simple compound **tesmilifene** (**133**) has been shown to enhance the antitumor activity of several classes of chemotherapy agents against multidrug resistant cancers. The mechanism of action of this drug is as yet unclear. This compound is prepared by alkylation of *p*-benzylphenol (**131**) with 2-chlorotriethylamine (**132**).[21]

Moving the side chain bearing the amino group to the benzylic carbon leads to compounds with very different activities. Anticholinergic agents specific for muscarinic receptors have proven to be very useful for treating urinary incontinence. The synthesis of a recent example speculatively starts with the Fries rearrangement of the ester (**134**) to the benzophenone (**135**). This compound is then treated with the organometallic reagent obtained from 2-chloroethyl diidopropyl amine (**136**) to afford the benzhydrol (**137**). Hydrogenolysis over palladium would then afford **tolterodine** (**138**).[22]

The release of large amounts of the neurotransmitter glutamate during a stroke lead to flood nerve cell receptors, such as that for N-methyl-D-aspartate (NMDA) causing excessive stimulation. The NMDA receptor antagonists have as a result been investigated as agents for preventing some of the injury from strokes. Relatively small changes in the structure of tolterodine, lead to an NMDA antagonist. Condensation of 3,3-difluoro-benzophenone (**139**) with the lithio reagent from acetonitrile yields the benzydrol (**140**). The nitrile group is then reduced to the corresponding amine by means of diborane. Treatment with acid leads to loss of the tertiary hydroxyl group and formation of the olefin (**141**). The double bond is reduced by catalytic hydrogenation (**142**).[23] The primary amino group can then be converted to the N-methyl derivative by any of a number of procedures, such as, treatment with formaldehyde and formic acid. Then the NMDA receptor antagonist **delucemine** (**143**) is obtained.

The first product from one of the branches of the arachidonic acid cascade comprises a bridged bicyclic compound where a pair of oxygen atoms form one of the rings. One route from this leads to the familiar five-membered ring prostaglandins; an alternate path leads to thromboxanes, one of the principal injurious products of the cascade. Compounds that block thromboxanes would be useful in treating the vaso-constricting and platelet aggregating action of this compound. The synthesis of a recent thromboxanes A_2 receptor antagonist starts with the Wittig condensation of the ylide from ω-phosphinocarboxylic acid (**145**) with the ketone (**144**) to give the olefin (**146**). The reaction apparently proceeds to give exclusively to the (E) olefin isomer. Acid hydrolysis removes the acetyl protection. Reaction of that intermediate with the cyanoimidate **148** leads to displacement of one phenoxy groups by the amine in **147**. This reagent may be viewed formally as the phenol acetal of cyanamide formate. Reaction of the intermediate **149** with *tert*-butylamine displaces the remaining phenoxy group to form the corresponding cyanoguanidine. Thus **terbogrel (150)** is obtained.[24]

6. MISCELLANEOUS MONOCYCLIC AROMATIC COMPOUNDS

Damage to peripheral nerves, termed peripheral neuropathy, is among one of the more common consequences of diabetes; it may also be caused by Parkinson's disease or various toxins. Neurophilins are agents that reverse that condition and encourage new nerve growth. The small molecule **timcodar (157)** has shown promising activity in animal models and several early clinical trials. Aldol condensation of pyridine-4-carboxaldehyde **(152)** with acetone dicarboxylic acid **(151)** proceeds to the keto acid **(153)**. This transient intermediate decarboxylates under reaction conditions to afford the bis(enone) **154**. Hydrogenation over platinum proceeds to give the saturated ketone **(155)**. Treatment of **155** with benzylamine in the presence of cyanoborohydride leads to the product from reductive amination **(156)**. Acylation with hydrocinnamoyl chloride affords the amide **157**.[25]

The discovery several decades ago of the nontricyclic antidepressant drugs, exemplified by Prozac (fluoxetine), has revolutionized society's view of depression. Use of these agents has mushroomed to the point where they are indiscriminately consumed to overcome moods induced by mild disappointments. The enormous market for drugs that act by this mechanism has led to the introduction of a host of more or less closely structurally analogues. A very recent stereoselective synthesis for one of these drugs, (*S,S*)**reboxetine** (**166**), starts with the commercially available chiral (*S*)-3-aminopropanediol (**158**). Acylation with chloroacetyl chloride leads to the amide (**159**). Treatment of intermediate **159** with strong base result in internal displacement of halogen with consequent formation of the morpholine ring (**160**). Reduction of the amide function with the hydride Red-Al forms the desired morpholine (**161**). The secondary amino group is protected as its *t*-BOC derivative (**162**) by acylation with *t*-BOC chloride. The next step involves oxidation of the primary alcohol with the unusual reagent combination consisting of 2,2,6,6-tetramethylpiperidinyl-*N*-oxide (TEMPO) and trichloroisocyanuryl chloride. Thus aldehyde **163** is obtained. Condensation of this with diphenyl zinc, obtained by treating phenylmagnesium bromide with zinc bromide, affords the secondary carbinol (**164**). The same reaction in the absence of zinc leads to recovery of unreacted aldehyde. The desired diastereomers was formed in an ~3:1 ratio with its isomer. The final piece could be added by conventional means, for example, by reaction with 2-ethoxyphenol in the presence of DEAD and carbon tetrachloride. Reaction of **104** with the chromyl reagent (**165**) followed by oxidative removal of chromium by iodine gave the product in high yield. Removal of the *t*-BOC protecting group with trifluoroacetic acid (TFA) completes the synthesis of (**166**).[26]

Many, if not most, cancer chemotherapy agents can be considered selective toxins with a rather narrow therapeutic index. The administration of these agents is quite frequently accompanied by nausea and vomiting, the reflex whereby the body seeks to reject toxins. Classical antiemetic compounds, such as the phenothiazines, have little effect on chemotherapy induced emesis. The more recently introduced serotonin antagonists are far more effective. They however, are, not universally effective and elicit severe CNS side effects in some patients. A pair of closely related antiemetics act by a very different mechanism: these drugs oppose emesis by acting as antagonists at tachykinin receptors. The synthesis of the common moiety starts with aldol-like condensation of the anion from the acetonitrile (**168**) generated with a strong base with the ester group on the piperidine (**167**). Heating the product (**169**) from that reaction with strong acid leads to hydrolysis of the nitrile group to the corresponding acid. The transient intermediate then decarboxylates to afford the ketone (**170**). Reaction of **170** with bromine then gives the bromoketone (**171**). This undergoes spontaneous internal displacement of bromine by the piperidine nitrogen. The formation of this bridge leads to the quinuclidine (**172**) as a quaternary salt. The benzyl protecting group on nitrogen is then removed by hydrogenolysis over palladium to yield the substituted quinuclidone (**173**). Reductive amination with 2-methoxy-4-isopropylbenzylamine (**174**) affords **ezlopipant** (**175**); the same reaction using 2-methoxy-4-*tert*-butylylbenzylamine leads to **maropitant (176)**.[27,28]

Most paradigms for treating Parkinson's disease involve increasing dopamine levels at synapses in the brain. Administration of dopamine itself is ruled out since the polarity of this compound prevents it from crossing the blood–brain barrier. Dopamine, like most neurotransmitters, is taken back by the presynaptic fiber after it has triggered action. A compound that inhibits this process would as a consequence increase levels of dopamine in the synaptic cleft. The starting material (**177**) for a dopamine reuptake inhibitor in fact comprises the free acid form of cocaine. Hydrolysis in mild acid serves to remove the benzoate (**178**). Reaction of this hydroxyl acid with phosphorus oxychloride leads to loss of the hydroxyl and formation of the conjugated acid. This compound is then converted to its methyl ester (**179**). Condensation with the Grignard reagent from 3,4-dichlorobromobenzene in the presence of copper leads to conjugate addition of the organometalic reagent and formation of **180**. The ester grouping is next reduced to the corresponding carbinol (**181**). Swern oxidation with oxalyl chloride then leads to aldehyde **182**. Treatment of this intermediate with *O*-methylhydroxylamine leads to the *O*-methlyloxime. Thus the dopamine reuptake inhibitor **brasofensine (183)** is obtained.[29]

Both the older tricyclic antidepressants and the more recent drugs related to fluoxetine owe their efficacy to interaction with receptors for the neurotransmitters epinephrine, serotonin, and dopamine. The antidepressant **igemsine (191)** acts by some other as yet undefined mechamism. The compound is described as a ligand for sigma receptors, a subclass of opiate receptors not associated with pain pathways. Alkylation of the anion from 2-phenylbutyric acid with the allylic halide (**185**) yields the acid (**186**). The carboxylic acid is then converted to the corresponding azide by reaction in turn with thionyl chloride and sodium azide. Heating this compound in an aprotic solvent then affords the isocyanate rearrangement product. Reduction with LiAlH$_4$ gives the N-methylamine **189**. This compound is then acylated with the acid chloride from cyclopropyl carboxylic acid (**190**). Reduction of the amide, again with hydride, gives the antidepressant **191**.[30]

Virtually all estrogen antagonists that have been in the clinic incorporate a basic side chain in the form of an tertiary-aminoethoxy aromatic ether. An exception, published in the 1960s, reported that the amine could be replaced by a glycol.[31] The estrogen antagonist **ospemifene (195)** takes this one step further. Antiestrogenic activity is retained when the basic ether in the antagonist toremifene (**196**)[32] is replaced by a simple hydroxyethyl group. This agent is prepared starting with an intermediate (**192**) used to prepare **196**. Alkylation of the phenol in **192** with the choroethanol benzyl ether affords **193**. The free hydroxyl at the end of the chain is then converted to the halide (**194**) by reaction with carbon tetrachloride and triphenyl phosphine. Removal of the benzyl group by means of hydrogen over palladium then affords **195**.[33]

There is some evidence that points to an association between Alzheimer's disease and a defict in brain acetylcholine levels. Considerable attention has thus focused on cholinesterase inhibitors as potential drugs for treating the affliction. The preparation of a relatively simple inhibitor is prepared by acylation of the benzylamine (**197**) with chloroformamide (**198**). The resulting urethane is then resolved to afford **rivastigmine (199)**.[34]

The lipidemic compound **ezetimibe (207)**, whose structure differs markedly from the ACAT inhibitor avasamibe (**125**), discussed previously, also inhibits absorption of cholesterol from the gut. The key to the construction of this compound involves formation of an azetidone. Enamine formation between the *p*-benzyloxybenzaldehyde (**200**) and aniline (**201**) provides one of the required reactants, imine **202**. This compound is then treated with a half-acid chloride of ethyl adipate (**203**) and triethylamine. In all likelihood, this first dehydrohalogenates under reaction conditions to form the substituted ketene. The transient intermediate reacts with the imine in a 2 + 2 cylcoaddition to afford a four-membered ring and thus **204**. The reaction proceeds to give the trans isomer almost exclusively. The ester group is then hydrolyzed by means of lithium hydroxide. Condensation with the zinc reagent formed *in situ* from *p*-fluoromagnesium bromide and zinc chloride affords the ketone (**205**). The carbonyl group is then reduced with diborane to afford the alcohol (**206**). Removal of the benzyl protecting group by hydrogenolysis over palladium finally affords **207**.[35]

REFERENCES

1. H. Amschler, U.S. Patent 5,712,298 (1998).

2. K. Tsutsumi, T. Sagimoto, Y. Tsuda, E. Uesaak, K. Shinomiya, Y. Shoji, A. Shima, U.S. Patent 5,081,112 (1989).

3. R.S. Randad, M.A. Mahran, A.S. Mehanna, D.J. Abraham, *J. Med. Chem.* **34**, 752 (1991).

4. B.R. Henke et al., *J. Med. Chem.* **41**, 5020 (1998).

5. P.V. Devasthale et al., *J. Med. Chem.* **48**, 2248 (2005).

6. Anon U.S. Application 1992-903691.

7. L.M. Coussens, B. Fingleton, L.M. Matrisian, *Science* **295**, 2387 (2002).

8. H.C.E. Kluender et al., U.S. Patent 5,789,434 (1998).

9. J.S. Sawyer et al., *J. Med. Chem.* **38**, 4411 (1995).

10. K.H. Donaldson, B.G. Shearer, D.E. Uehling, U.S. Patent 6,251,925 (2001).

11. J. Malm, *Curr. Pharm. Design* **10**, 3525 (2004).

12. N. Yokoyama et al., *J. Med. Chem.* **38**, 695 (1995).

13. K.S. Atwal, U.S. Patent 6,013,668 (2000).

14. R.L. Cournoyer, P.F. Keitz, C. O'Yang, D.M. Yasuda, U.S. Patent 6,057,349 (2000).

15. L.A. Sobrera, P.A. Leeson, J.A. Castaner, M. del Fresno, *Drugs Future* **29**, 240 (2002).

16. K. Imaki, Y. Arai, T. Okegawa, U.S. Patent 5,017,610 (1991).

17. L. Bender, M.J. Melnick, U.S. Patent 5,753,653 (1998).

18. N. Murugesan et al., *J. Med. Chem.* **46**, 125 (2003).

19. H.T. Lee et al., *J. Med. Chem.* **39**, 5031 (1998).

20. H.-J. Lang, A. Weichert, H.-W. Kleeman, H. Englert, W. Scholtz, U. Albus, U.S. Patent 5,591,754 (1997).

21. L.J. Brandes, M.W. Hermonat, U.S. Patent 4,803,227 (1989).

22. N.A. Johnsson, B.A. Sparf, L. Mikiver, P. Moses, L. Nilvebrant, G. Glas, U.S. Patent 5,382,600 (1995).

23. S.T. Moe, D.L. Smith, K. By, J.A. Egan, C.N. Filer, *J. Label. Comp. Radiopharm.* **41**, 535 (1998).

24. R. Soyka, B.D. Guth, H.M. Weisenberger, P. Luger, T.H. Muller, *J. Med. Chem.* **42**, 1235 (1999).

25. R. Zelle, M. Su, U.S. Patent 5,840,736 (1998).

26. E. Brenner, R.M. Baldwin, G. Tramagnan, *Org. Lett.* **7**, 937 (2005).

27. J.A Lowe, U.S. Patent 5,451,586 (1995).

28. I. Fumitaka, H. Kondo, M. Nakane, K. Shimada, J.A. Lowe, T.J. Rosen, U.S. Patent 5,807,867 (1998).

29. L.A. Sobrera, J.A. Castaner, *Drugs Future* **25**, 196 (2000).

30. G.G. Aubard et al., U.S. Patent 5,034,419 (1991).

31. D. Lednicer, D.E. Emmert, S.C. Lyster, G.W. Duncan, *J. Med. Chem.* **12**, 881 (1969).

32. D. Lednicer, *The Organic Chemistry of Drug Synthesis*, Vol. 5, John Wiley & Sons, Inc., NY, 1995, p. 33.

33. M. DeGregorio, V. Wiebe, L. Kangas, P. Harkonen, K. Vaananen, A. Laine, U.S. Patent 5,750,576 (1998).

34. R. Amstuz, M. Marzi, M. Boelsterli, M. Walsinshaw, *Helv. Chim. Acta* **73**, 739 (1990).

35. W.D. Vaccaro, R. Sher, H.R. Davis, *Bioorg. Med. Chem.* **6**, 1429 (1998).

CHAPTER 4

CARBOCYCLIC COMPOUNDS FUSED TO A BENZENE RING

The nucleus of a modest number of new compounds comprise a two- or three-ring fused system, one of which consists of a benzene ring. As was the case for free-standing benzene rings in Chapter 3, the annelated rings in most instances serve merely as supports for the pharmacophoric substituents.

1. INDENES

Imidazolines have a venerable history as α-adrenergic agents. Compounds that include this group variously act as α_1- and α_2-agonists or antagonists depending on the substitution pattern in the rest of the molecule. The indene **fadolmidine** (**5**), is an effective α_2-agonist that blocks pain responses. The compound does not cross the blood–brain barrier as a result of its hydrophobic character; it has as a consequence been developed as a drug for use as a spinal analgesic. Preparation of the compound starts with a crossed version of the McMurray reaction. Thus treatment of a mixture of the indanone (**1**) with *N*-benzyl protected imidazolecarbox-aldehyde (**2**) in the presence of TiCl$_2$, preformed from TiCl$_4$ and zinc powder, gives the coupling product **3**. Catalytic hydrogenation serves to

The Organic Chemistry of Drug Synthesis, Volume 7. By Daniel Lednicer
Copyright © 2008 John Wiley & Sons, Inc.

both reduce the double bond and to remove the benzyl protecting group (**4**). Reaction of **4** with hydrogen bromide then cleaves the methyl ether on benzene to afford the free phenol and **5**.[1]

The antidepressant compound **lubazodone** (**8**) illustrates the breadth of the structural requirements for serotonin selective reuptake inhibitors; the structure of this agent departs markedly from that of fluoxetine, the first drug in this class. The compound at hand also exemplifies the current trend for preparing drugs in chiral form. Thus reaction of the indanol (**6**) with the mesylate from chiral glycidic oxide in the presence of base leads to the epoxypropyl ether (**7**) with retention of chirality. Treatment intermediate **7** with aminoethylsulfonic acid closes the morpholine ring. Product **8** consists of pure (*S*) enantiomer.[2]

Alzheimer's disease, as noted earlier, is associated with decreased levels of acetyl choline in the brain. Most of the drugs that have been introduced to date for treating this disease thus comprise anticholinergic agents intended to raise the deficient levels by inhibiting loss of existing acetyl-choline. A compound-based on an indene perhaps surprisingly, shows anticholinergic activity, and has been proposed for treatment of Alzheimer's disease. Condensation of piperidine aldehyde (**10**) with the indanone (**9**) leads to the olefin (**11**). Catalytic reduction removes the double bond to afford **donepezil** (**12**).[3]

More recent work indicates that monoamine oxidase (MAO) inhibitors may be useful as well. The indene **ladostigil** (**17**) is intended to address both of those targets; the compound thus incorporates a carbamate group associated with anticholineric activity and a propargyl moiety found in MAO inhibitors. The synthesis involves juggling protecting groups on two reactive functions. Thus, reaction of amino-indanol (**13**) with bis-*tert*-butoxy carbonate affords the corresponding *t*-BOC protected derivative **14**. Treatment of derivative **14** with *N*-methy-*N*-ethyl carbamoyl chloride affords the *O*-acylated carbamate (**15**). The *t*-BOC protecting group is then removed by means of hydrogen chloride to give the free amine (**16**). Reaction of **16** with propargyl bromide gives **17**. This drug also consists of a single (*R*) enantiomer; it is not clear from the source[4] at which stage the resolution takes place.

The hormone melatonin (**30**) is intimately involved in the diurnal cycle with levels rising late in the day prior to sleep. Congeners have as a result been prepared in the search for a sleep inducing drugs. **Ramelteon (29)**, an indene that incorporates several structural features of the hormone, has been approved by the FDA as a sleeping aid. The first part of the synthesis involves construction of the ethylamine side chain. Thus condensation of indanone (**18**) with the yilde from 2-diethoxyphoshonoacetonitrile attaches the requisite two carbon chain (**19**). The nitrile is then reduced to the corresponding primary amine (**20**) by means of Raney nickel. This drug also follows the current trend toward a chiraly defined substance. Reduction of the double bond with rhodium in the presence of the chiral catalyst 2,2′-bis(diphenylphosphino)-1,1′-binaphthyl (BiNAP) affords the intermediate (**21**) as the (*S*) enantiomer. This compound is then acylated with propionyl chloride to afford **22**. The remainder of the scheme involves construction of the fused furan ring. Bromination proceeds at the slightly less hindered position. The methoxy group is then cleaved with boron tribromide to yield the bromophenol (**23**). Alkylation of the phenol with allyl bromide proceeds to the allyl ether (**24**). Heating compound **24** leads to a Claisen rearrangement to the *C*-allyl derivative (**25**); bromine at the other ortho position prevents formation of the alternate undesired isomer. Ozonization followed by reductive workup leads to **26**; treatment of the aldehyde with sodium borohydride reduces that function to an alcohol providing **27**. The blocking bromine atom is then replaced by hydrogen by means of hydrogenation over palladium to yield **27**. Construction of the furan ring starts by conversion of the primary alcohol to its mesylate with methanesulfonyl chloride. Treatment of the product with triethylamine (TEA) forms the phenoxide, which then displaces the mesylate group. This internal displacement forms the furan ring. Thus, ramelteon (**29**) is obtained.[5]

2. NAPHTHALENES

The parathyroid glands comprise one of the principal centers for regulation of calcium levels. The parathyroid hormone secreted by those glands into the bloodstream directly controls levels of calcium and phosphorus. In the normal course of events, low levels of calcium will result in release of the hormone and vice versa. Patients with kidney disease who are on dialysis, as well as those with parathyroid gland neoplasms, tend to have significantly elevated calcium levels, a condition that can lead to hypertension and congestive heart failure. A structurally relatively simple naphthalene derivative lowers parathyroid levels by binding to calcium receptors on the gland. Reaction of commercially available (*R*)-ethyl-α-naphthyl amine (**31**) with the aldehyde (**32**) affords the Schiff base (**33**). Reduction of intermediate **33** with cyanoborohydride leads to the calcium mimetic compound **cinalcet** (**34**).[6]

It is now recognized that protein kinases play an important role in intra- and intercellular communications. These enzymes are consequently directly involved in cell proliferation. The p38 kinase, for example, regulates the production of key inflammatory mediators. Excess expression of this factor is involved in the pathology of rheumatoid arthritis, psoriasis, and Crohn's disease. A rather complex protein kinase inhibitor, which includes a substituted naphthyl moiety, has shown preliminary *in vivo* activity. The convergent synthesis starts with construction of a heterocyclic fragment. Condensation of the keto-nitrile (**35**) with *p*-tolylhydrazine (**36**) proceeds to give the pyrazole (**37**). The overall transform can be rationalized by initial formation of a hydrazone; addition of the remaining hydrazine nitrogen to the nitrile would then form the pyrazole ring. Reaction of this intermediate with phosgene then converts the primary amine to an isocyanate (**38**). The other branch of the scheme first involves alkylation of the *t*-BOC protected naphthylamine (**39**) in the presence of base with chloroethyl morpholine (**40**). Exposure to acid then cleaves the *t*-BOC group to afford the free amine (**41**). Addition of the amino group in this intermediate to the reactive iscocyanate in **38** connects the two halves via a newly formed urea function. Thus, the p38 kinase inhibitor **doramapimod (42)** is obtained.[7]

3. TETRAHYDRONAPHTHALENES

Drug therapy for treating Parkinson's disease involves supplementing the deficient levels of dopamine in the brain that characterize the disease. The blood–brain barrier virtually dictates that the agents need to be lipophillic;

dopamine itself is too hydrophilic to penetrate the brain from the circulation. A tetrahydronaphthalene derivative that incorporates one of the phenolic hydroxyls, as well as an ethylamine-like sequence of dopamine, shows the same activity as the neurotransmitter. The lipophilicity of **rotigotine (52)** allows it not only to cross the blood–brain barrier, but to also reach the circulation when administered topically. The drug is in fact provided in a skin patch formulation that provides prolonged blood levels. The preparation of this dopaminergic agent begins with the conversion of the dihydroxynaphthalene (**43**) to its methyl ether by means of dimethyl sulfate. Treatment of **44** with sodium in alcohol initially leads to the dihydro intermediate (**45**). The regiochemistry follows from the fact that only the right-hand ring has two open positions in a 1,4 relationship to the methyl ether for forming the initial metal adduct. Treatment of **45** with acid then hydrolyzes the enol ether to afford the β-tetralone (**46**). The carbonyl group is next converted to a Schiff base (**47**) by reaction with propylamine. Catalytic hydrogenation of the intermediate then affords the secondary amine. This intermediate is resolved via its dibenzoyl tartrate salt (**48**). The methyl ether in the (*S*) enantiomer is cleaved by means of hydrogen bromide to give the corresponding phenol (**49**). Reaction with an activated form of 2-thienylacetic acid (**50**) followed by reduction of the amide (**51**) with diborane gives the corresponding tertiary amine, **52**.[8]

The effects on cell proliferation of retinoids has led to the investigation of structurally related compounds as potential antineoplastic drugs. The finding many years ago that the cyclobexyl moiety in the naturally occurring compound can be replaced by a tetrahydronapthalene has simplified work in this area. It is of interest that in each of the examples that follow a benzene ring serves as a surrogate for the unsaturated carbon chain found in natural retinoids. The tetralin-based compound **tamibarotene** (**59**) has been tested as an agent for treating leukemias. Reaction of the diol (**53**) with hydrogen chloride affords the corresponding dichloro derivative (**54**). Aluminum chloride mediated Friedel–Crafts alkylation of acetanilide with the dichloride affords the tetralin (**55**). Basic hydrolysis leads to the primary amine (**56**). Acylation of the primary amino group with the half acid chloride half ester from terephthalic acid (**57**) leads to the amide (**58**). Basic hydrolysis of the ester grouping then affords **59**.[9]

The retinoid-like compound **bexarotene** (**63**) is approved for treating skin lesions associated with T-cell lymphomas. The starting tetralin (**60**) is probably obtained by alkylation of toluene with dichloride (**54**). Friedel–Crafts acylation with the acid chloride (**57**), gives the ketone (**61**). This intermediate is then treated with the ylide from triphenylmethylphosphonium bromide. The carbonyl oxygen in the product (**62**) is now replaced by a methylene group. Saponfication of the ester affords the free acid and thus **63**.[10]

The saga of estrogen antagonists had its start in the mid-1960s when several series of compound related to diethylstilbestrol were investigated as nonsteroidal antifertility agents. The early work culminated in the development of tamoxifen, an estrogen antagonist that was and still is widely used as adjuvant treatment for breast cancer. The benzothiophene-based analogue raloxifene, which retains some estrogenic activity, was introduced more recently as a drug for treating osteoporosis. Other more recent examples will be found scattered throughout this volume. A tetralin ring forms the nucleus of another recent entry, **lasofoxifene (74)**. One synthesis for the penultimate intermediate starts with the formylation of the desoxybenzoin (**64**) with ethyl formate and sodium ethoxide to its sodium salt (**65**). In a convergent step, the benzyl chloride (**66**) is allowed to react with triphenylphosphine to give the corresponding phosphonium salt (**67**). Reaction of **67** with the salt (**65**) leads directly to the product from coupling with the ylide as a mixture of isomers. This mixture is then hydrogenated to give the ketone (**68**). Treatment of **68** with 3 equiv of aluminum chloride results in scission of the methyl ether on the most electron-deficient ether to give the phenol (**69**). This compound is then cyclized to the corresponding dihydronaphthalene with toluenesulfonic acid (TSA). The basic ether side chain required for antagonist activity is added by alkylation with 2-chloroethyl pyrrolidine to afford, nafoxidine (**72**). Catalytic hydrogenation of this product gives the tetralin (**73**).[11] Reaction of **73** with boron tribromide results in cleavage of the methyl ether in the fused ring to give (**74**).[12]

The more recent synthesis for lasoxifene (**74**) takes a very different course. The first step comprises displacement of one of the halogens in 1,4-dibromo-benzene by the alkoxide from *N*-2-hydroxyethylpyrrolidine **75** in the presence of 18-crown ether to afford **76**. Condensation of the lithium salt from **76** with 6-methoxytetralone (**77**) followed by dehydration of the initially formed carbinol yields intermediate **78**, which incorporates the important basic ether. Reaction of this compound with pridinium bromide perbromide leads to displacement of the vinylic proton by halogen and formation of the bromide **79**. Condensation of this product with phenylboronic acid in the presence of a palladium catalyst leads to coupling of the phenyl group by formal displacement of bromine. The product (**79**), is then taken on to **74** as above.[12]

4. OTHER BENZOFUSED CARBOCYCLIC COMPOUNDS

Multidrug resistance, as noted earlier, is the all too prevalent phenomenon where a patient's resistance to one class of cancer chemotherapy agents comes to encompass mechanistically quite different drugs. Compounds with a wide variety of structural features have shown at least preliminary activity in resolving this problem. The structurally rather complex agent **zosuquidar** (**87**) has shown promising activity against this problem. Reaction of dibenzosuberone (**80**) with the difluorocarbene from chlorodifluoroacetate affords the cyclopropyl adduct (**81**). Reduction of the ketone with borohydride proceeds to afford the derivative wherein the fused cyclpropyl and alcohol are on the same side of the seven-membered ring.

The carbinol (**82**) is then converted to the halide with thionyl chloride apparently with retention of configuration (**83**). Displacement with piperazine monoformamide leads to the alkylated product in which the groups are now anti. Hydrolysis of the formamide grouping then affords secondary amine **84**. In a convergent sequence, 5-hydroxyquinoline (**85**) is allowed to react with the tosyl derivative of chiral glycidol. The epoxy group in the product (**86**) retains configuration. Condensation of piperazine (**84**) with the quinoline (**86**) opens the epoxide to afford **87**. Configuration of the alcohol is again retained as the reaction takes place at the nonchiral terminal of the side-chain to be. Thus, **87** is obtained.[13]

Most of the products from the arachidonic acid cascade exert deleterious effects. Prostacylin (**100**) is a notable exception because of its vasodilatary activity. The compound is destroyed far too quickly to be used as a drug. An analogue in which a fused tetralin moiety replaces the furan and part of the side chain is approved for use as a vasodilator for use in patients with pulmonary hypertension. The lengthy complex synthesis starts with the protection of the hydroxyl group in benzyl alcohol (**88**) by reaction with *tert*-butyl dimethyl silyl (TMBDS) chloride (**89**). Alkylation of the anion from **89** (butyllithium) with ally bromide affords **90**. The protecting group is then removed and the benzylic hydroxyl is oxidized with oxalyl chloride in the presence of triethylamine to give the benzaldehyde (**91**). The carbonyl group is then condensed with the organomagnesium derivative from treatment of chiral acetylene (**92**) with ethyl Grignard to afford **93**. (The triple bond is not depicted in true linear form to simplify the scheme.) The next few steps adjust the stereochemistry of the newly formed alcohol in **93**. This group is first oxidized back to a ketone with pyridinium chlorochromate. Reduction with diborane in the presence of chiral 2-(hydroxyl-diphenylmethyl)pyrolidine affords the alcohol as a single enantiomer. This compound is then again protected as its TMBDS ether. Heating **94** with cobalt carbonyl leads to formation of the tricyclic ring system (**95**). Mechanistic considerations aside, the overall sequence to the product (**95**) involves eletrocylic formation of the six-membered ring from the olefin and acetylenic bond, as well as insertion of the elements of carbon monoxide to form the five-membered ring. Catalytic hydrogenation of **95** leads to reduction of the double bond in the enone, as well as hydrogenolysis of the benzylic TMBDS ether on the six-membered ring (**96**). Reduction of the ketone then leads to the alcohol apparently as a single enantiomer. Acid leads to loss of the tetrahydropyrany protecting group to afford intermediate **97**. The presence of labile groups in this compound precludes the usual methods, such as hydrogen bromide or boron tribromide, for cleaving the methyl ether. Instead, in an unusual sequence, the phenol (**98**) is obtained by treatment of **97** with butyllithium and diphenylphosphine. The product (**98**) is then alkylated with 2-chloroacetonitrile. Hydrolysis of the cyano group to an acid finally affords the vasodilator **treprostinil (99)**.[14–16]

REFERENCES

1. A. Karjalainen, P. Huhtala, S. Wurster, M. Eloranta, M. Hillila, R. Saxlund, V. Cockroft, A. Karjalainen, U.S. Patent 6,313,322 B1 (2001).

2. M. Fuji, T. Suzuki, S. Hayashibe, S. Tsukamoto, S. Yatsugi, T. Yamaguchi, U.S. Patent 5,521,180 (1996).

3. Y. Imura, M. Mishima, M. Sugimoto, *J. Label. Comp. Radiopharm* **27**, 835 (1989).

4. M. Chorev, T. Goren, Y. Herzig, J. Sterling, M. Weinstock-Rosin, M.B.H. Youdim, U.S. Patent 6,462,222 (2002).

5. K. Chilma_Blair, J. Castaner, J.S. Silvestre, H. Bayes, *Drugs Future* **28**, 950 (2003).

6. B.C. Van Wagenen, S.T. Moe, M.F. Balandrin, E.G. DelMar, E.F. Nemeth, U.S. Patent 6,211,244 (2001).

7. J. Regan et al., *J. Med. Chem.* **45**, 2994 (2002).

8. N.J. Cusack, J.V. Peck, *Drugs Future* **18**, 1005 (1993).

9. Y. Hamada, I. Yamada, M. Uenaka, T. Sakata, U.S. Patent 5,214,202 (1993).

10. M.F. Boehm, R.A. Heyman, L. Zhi, C.K. Hwang, S. White, A. Nadzan, U.S. Patent 5,780,676 (1998).

11. D. Lednicer, D.E. Emmert, S.C. Lyster, G.W. Duncan, *J. Med. Chem.* **12**, 881 (1969).

12. C.O. Cameron, P.A. Dasilva Jardine, R.L. Rosati, U.S. Patent 5,552,412 (1996).

13. J.R. Pfister et al., *Bioorg. Med. Chem. Lett.* **5**, 2473 (1995).

14. P.A. Aristoff, U.S. Patent 4,306,075 (1981).

15. P.A. Aristoff, U.S. Patent 4,649,689 (1982).

16. P.A. Aristoff, U.S. Patent 4,683,330 (1987).

CHAPTER 5

FIVE-MEMBERED HETEROCYCLES

The specific chapter to which a given drug is assigned is to some extent arbitrary. More than a few compounds in the preceding chapters included a heterocyclic ring in their structures. That fragment more often that not, however, comprised a cyclic base, for example, piperidine. The compound in question was not classified as a heterocycle as it is quite likely that it would show the same qualitative biological activity if that moiety was replaced by a noncyclic base. Heterocyclic moieties do, however, seem to play a role in the biological activity beyond simply providing a basic center for a good many agents. Compounds **7** and **18** provide a particularly apt example; the pyrrolidine ring in these enzyme inhibitors acts as a surrogate for a proline moiety that occurs in the natural substrate. Compounds meeting that criterion will be found in this and the following sections. Note that close to two-thirds of the compounds in this volume have been judged to meet that criterion and will be met in the following chapters.

1. COMPOUNDS WITH ONE HETEROATOM

The protease enzyme dipeptidal peptidase (DPP) is closely involved in glucose control. This enzyme regulates levels of the hormone-like

The Organic Chemistry of Drug Synthesis, Volume 7. By Daniel Lednicer
Copyright © 2008 John Wiley & Sons, Inc.

peptide incretin, which stimulates release of insulin by cleaving the molecule to an inactive form. Inhibition of DPP in effect extends the action of incretin. This helps prevent the increased levels of blood glucose that mark diabetes. The protease inhibitor **vidagliptin** (**7**), which is modeled in part on the terminal sequence in DPP, has been found to sustain levels of insulin in Type 2 diabetics. Inhibition is apparently reversible in spite of the presence in the structure of the relatively reactive α-aminonitrile function. Construction of one intermediate in the convergent synthesis comprises reaction of amino adamantamine (**1**) with a mixture of nitric and sulfuric acid. This reaction affords the product **2** from nitration of one of the remaining unsubstituted ternary positions. Treatment of this product with strong base leads to solvolysis of the nitro group to give aminoalcohol **3**. Preparation of the other moiety involves first acylation of the pyrrolidine (**4**) with chloroacetyl chloride to give amide **5**. Reaction of that intermediate with trifluoracetic anhydride (TFAA) converts the amide at the 2 position to the correspounding nitrile. Alkylation of the adamantamine (**3**) with **6** proceeds on nitrogen to afford **7**.[1]

A substituted pyrrolidine, which acts as a DPP inhibitor, comprises another example in which this ring serves as a surrogate for proline. This compound is being investigated as an anticancer drug as a result of the finding that it inhibits growth of tumors in various animal models. The structure of this compound is notable for the rare occurrence of boron in the structure; in this case in the form of a covalently bound boronic acid. The final compound, **talabostat** (**18**), is comprised of a single enantiomer. This is accomplished in the case at hand by a stereoselective synthesis rather than by resolution of the final compound or an intermediate. The first step in the synthesis comprises protecting the amine in pyrrolidine (**8**) by conversion to its *tert*-butoxycarbonyl (**9**, *t*-BOC) derivative with *tert*-butoxycarbonyl anhydride. Reaction of the product with butyllithium generates an anion on carbon next to nitrogen. Treatment of this compound with triethyl borate displaces one of the ethoxy groups in the reagent to form a carbon–boron bond. The product is comprised of a 1:1 mixture of enantiomers. Hydrolysis of this intermediate then affords the corresponding boronic acid (**10**). A key step involves formation of the acetal-like compound of **12** with naturally occurring (+) pinanediol (**11**). The initial product is comprised of two diastereomers due to the fact that the starting boronic acid (**10**) consists of two enantiomers. The pair of diastereomers of **13** are then separated by recrystallization. In the next step, the desired isomer is coupled with the to t-BOC derivative from valine to give amide **16**. The pinane diol group is then removed by exchange with excess phenyl boronic acid. The final compound is converted to a salt (**18**) in order to avoid the formation of a stable zwitterion between the amine and the boronic acid function. Thus, **18** is obtained.[2]

As noted in Chapter 3, endothelins rank among the most potent known vasoconstricting agents; they have been implicated in a number of diseases including cerebral vasospasm and pulmonary hypertension. The stereoselective synthesis of an endothelin antagonist begins with the establishment of the chiral locus that will dictate the remaining asymmetric centers. Oxazolidone (**20**), derived from valine serves as the chiral auxiliary for that step. Condensation of the mixed-anhydride (**19**) from piperonalacetic acid, with the anion from auxiliary **20**, give the corresponding amide (**21**). Treatment of this intermediate with strong base followed by *tert*-butyl bromoacetate leads to the alkyladion product **22** as virtually a single isomer. The auxiliary heterocycle is then removed by means of lithium hydroperoxide to afford the half ester (**23**). Reaction of **23** with diborane selectively reduces the free acid to give the esteralcohol (**24**). The hydroxyl group is then activated by conversion to its tosylate (**25**). Treatment of that intermediate with anisyl hydroxylamine (**26**) in the presence of cesium carbonate affords the *O*-alkylated derivative **27**. The ester grouping is then exchanged with methylorthoformate to afford the methyl ester (**28**). Reaction of this last intermediate with trimethylsilyl triflate and butylamine in the presence of 1,2-dichloroethane presumably forms an anion-like species on the carbon adjacent to the ester. This then adds internally to the oxime carbon atom to yield a 1,2-oxazine. This product (**29**) predominates over the diastereomer in a 9 : 1 ratio.

Catalytic hydrogenation of **29** over palladium on charcoal results in scission of the weak N—O bond and formation of aminoalcohol **30**. This compound is converted to a pyrrolidine by an internal alkylation reaction. Thus, reaction of the intermediate with carbon tetrabromide and triphenyl phosphine presumably converts the alcohol to a bromide; internal displacement by the primary amine forms the five-membered ring. Alkylation of that amine with the complex bromo amide **32** affords the endothelin antagonist **atrasentan (33)**.[3]

Research on anticholinergic compounds has experienced something of a resurgence as a result of their utility in treating conditions such as urinary incontinence. The structures of these compounds are quite varied as shown by **darfenacin (44)**, which differs considerably from other compounds in this class, for example, tolterodine (Chapter 3). The synthesis of this compound (**44**) is also designed to produce a single enantiomer. Heat induced decarboxylation of proline (**34**) affords the key intermediate **35** as a pure enantiomer. The amino group is then converted to its tosylate (**36**) with tosyl chloride; the hydroxyl group interestingly does not react under those conditions. Converting that group to its derivative is accomplished by the Mitsonobu reaction with methyl tosylate to give the doubly derivatized intermediate **37**. Condensation of **37** with the anion from diphenyl acetonitrile (**38**), produced by reaction with sodium hydride, yields the alkylation product **39**. Treatment of this intermediate with hydrogen bromide removes the protecting group on nitrogen. The nitrile is then converted to the corresponding amide with sulfuric acid. In a converging scheme, acylation of benzofuran (**41**) with chloroacetyl chloride and aluminum chloride yields the chloroketone (**42**). Reaction of that with pyrrolidine (**40**) leads to the alkylation product **43**. Catalytic hydrogenation over palladium reduces the aryl carbonyl group to a methylene probably via the initially formed labile benzyl alcohol. Thus the anticholinergic agent **44** is obtained.[4]

Protein kinases comprise a series of closely related enzymes that catalyze the phosphorylation of hydroxyl groups in enzymes, which regulate a wide range of physiological processes. Phosphorylation may either turn the function of the target on or off. Inhibitors of the protein C kinases (PKC) that are involved in cell proliferation have attracted particular

Ruboxystaurin

attention as potential drugs. The complex fermentation product Staurosprin is a PKC inhibitor that has shown antineoplastic activity in a range of biological assays, but proved to be too toxic for use as a drug. The somewhat simpler analogue, ruboxystaurin shows greater potency and selectivity for specific PKCs. It has been pursued in the clinic to treat complications from diabetes. The synthetic acyclic product **enzastaurin (54)** is even more specific, inhibiting a subclass of PKCs involved in cell proliferation.

Reaction of the pyridyl methylamine (**45**) with methyl acrylate results in Michael addition of 2 equiv of the reagent to afford the diester (**46**).

Base-catalyzed Claisen condensation forms the desired six-membered ring. Heating that intermediate with base leads to saponification of the ester. The beta ketoacid decarboxylates under reaction conditions to form **47**.[5] Reductive amination of the carbonyl group in the ketone in this intermediate with aniline would then lead to the substituted aniline (**48**). Treatment of **48** with chloroacetontrile in the presence of boron trichloride leads to acylation on the benzene ring. The regiochemistry is attributable to the strongly acidic reaction conditions that inactivate the pyridine moiety. Aqueous workup then affords the chloroacetyl derivative (**49**). The ketone is then reduced by means of sodium borohydride. On heating, the product undergoes internal alkylation; the first formed compound then dehydrates to afford the substituted indole (**50**).[6] The next step in the scheme involves construction of the pyrrolodione moiety. The required carbon atoms are added by acylation of **50** with oxalyl chloride to afford **51**. This compound is then condensed with the imidate (**52**) from indole acetamide to afford an intermediate, such as **53**. The reaction may be visualized as involving first formation of an amide from imidate nitrogen with the acid chloride followed by addition of the anion from the methylene group in **52** to the carbonyl. The initially formed pyrroline is then dehydrated with strong acid.[7] Thus, the PKC inhibitor **54** is obtained.

Leukotrienes, products from one branch of the arachidonic cascade, are closely associated with symptoms of allergy, as well as asthma (*see* Chapter 3, etalocib). The benzothiophene-based leukotriene antagonist, **zileuton**, one of the first agents in this class, is now on the market. A related compound, **atreluton** (**60**), that omits the fused benzene ring present in the prototype, shows improved potency and duration of action over its predecessor. Condensation of benzyl bromide (**55**) with the anion from thiophene and butyllithium in the presence of the Heck catalyst [tetrakis(triphenylphosphine) palladium(0) gives the coupling product **56**. Reaction with NBS leads to the bromothiophene (**57**). Condensation of that intermediate with the methyl–ethynyl carbinol in the presence of triphenylphosphine, Heck catalyst, and cupric iodide leads to the coupling product **58**. The requisite functionality is constructed by first replacing the hydroxyl next to the acetylene by nitrogen. Mitsonobu-like reaction with *O,N* bis(phenyloxycarbonyl hydroxylamine) in the presence of triphenylphosphine and DEAD affords **59**. Reaction of this intermediate with ammonia leads to displacement of both phenoxy groups. This leads to formation of the free hydroxyl from the *O*-carbonate and a urea from the phenoxy ester, yielding the leukotriene antagonist **60**.[8]

2. COMPOUNDS WITH TWO HETEROATOMS

A. Oxazole and Isoxazoles

The discovery that non-steroid antiinflammatory agents (NSAID) owe their efficacy to inhibition of cyclooxygenase (COX), the enzyme that catalyzes the formation of prostaglandins was followed some time later by the finding that the enzyme occurred in several subtypes. The almost simultaneous discovery of specific inhibitors of COX-2, promised NSAIDS that reduced inflammation that spared the production of prostaglandins that maintain integrity of the stomach wall. The enormous success of the first agent on the market celecoxib, led to detailed investigation of the structure–activity relationship (SAR) of this class in competing laboratories. Minimum requirement for activity seemed to involve two aromatic rings on adjacent positions on a five-membered heterocyle. The very recent entry, **tilmacoxib (67)**, shows that one of those benzene rings can be replaced by cyclohexane. Condensation of *m*-fluorobenzyl bromide (**62**) with the acid chloride (**61**) from cyclohexane carboxylic acid in the presence of the Heck reagent affords the ketone (**63**) that incorporates the requisite two rings. Bromination proceeds on the benzylic position to afford **64**. This reactive halogen is displaced with acetate to give the key intermediate (**65**). Construction of the heterocyclic ring involves reaction of **65** with ammonium acetate. The reaction can be viewed for bookkeeping considerations as proceeding through the initial reaction of ammonium ion with the ketone to form an imine. This imine cyclizes with the adjacent carbonyl on the acetate to afford **66**. Treatment of **66** with chlorosulfonic acid gives the sulfonyl chloride; that sulfonyl chloride is immediately allowed to react with ammonia to yield the corresponding sulfonamide. This compound then affords the COX-2 inhibitor **67**.[9]

The large number of compounds that have been reported illustrate the wide selection of heterocyclic five-membered rings that are compatible with COX-2 inhibitory activity. The oxazole ring can, for example, be replaced by an isoxazole. Reaction of deoxybezoin (**68**) with hydroxylamine affords the oxime (**69**). Treatment of this intermediate with 2 equiv of butyllithium followed by acetic anhydride goes to the hydroxyl isoxazoline (**71**). The transform can be rationalized by assuming that this proceeds via the *O*-acylated intermediate (**70**). The anion from the benzylic position would add to the acetyl carbonyl group to afford the observed product. Reaction of this compound with chlorosulfonic acid results in sulfonation of the aromatic ring nearest nitrogen. The first step probably comprises dehydration of tertiary alcohol in **71** under the strongly acidic conditions to give **72**. Subsequent addition of ammonia converts the chlorosulfonic acid to the corresponding sulfonamide and thus **valdecoxib** (**73**).[10] Reaction of **73** with acetic anhydride leads to acylation of the amide nitrogen, which increases the acidity of that already acidic function. Treatment with base affords the water soluble salt **parecoxib** (**74**),[11] which is suitable for use in injectable formulations.

The immune system is believed to play a role in rheumatoid arthritis, in contrast to the far more common osteoarthritis, which is related to aging. An agent that has immunosuppressive action has proven useful for treating rheumatoid arthritis. This compound **lenflunomide (77)**, can be prepared in a single step by acylation of *p*-trifluoromethylaniline (**76**) with the commercially available isoxazole (**75**).[12] The isoxazole ring, it was later found, is readily cleaved *in vivo* to give the cyano ketone, **teriflunomide (78)**. Lenfluomides is consequently considered to be a prodrug for the latter. This agent can be prepared by condensation of the sodium salt from cyanoacetone (**79**) with *p*-trifluoromethylphenyl isocyanate (**80**),[13] itself readily accessible from the aniline (**76**). Teriflunomide, is used in the clinic as a potential drug for relieving some of the effects of multiple scelerosis.

B. Imidazoles and a Pyrrazole

By now, it is well established that "conazole" antifungal agents attack fungi by inhibiting the synthesis of steroids essential to the fungal life cycle. Virtually every antifungal agent in this class incorporates an imidazole ring into its structure. This moiety is thus, not surprisingly, found in the form of an enamine in a recent conazole. The starting material for this agent (**83**), could, for example, be prepared by bromination of the propiophenone (**81**) followed by displacement of halogen with imidazole. Alkylation of the enolate from the ketone with the side-chain fragment (**84**), yields the antifungal agents **omoconazole** (**85**).[14]

81; R = H
82; R = Br
83
84
NaH
85

Histamine H_3 receptors have been found to modulate the release of neurotransmitters, such as acetyl choline, dopamine, and serotonin, involved in alertness and cognitive function. Compounds that act as antagonists at those sites favor release of those neurotransmitters and result in increased alertness in animal models. Antagonists would hold promise for treatment of attention deficit syndrome and related conditions including even possibly Alzheimer's disease. The first step in the synthesis toward the antagonist **cipralisant** (**92**) comprises separation of the enantiomers of the carboxylic acid **86**. To this end, the acid is reacted with a chiral sultam derived from camphor. The resulting diastereomers (**87**) are then separated by chromatography. Each of the diastereomerically pure derivatives, only one of which is shown, is then treated in the cold with DIBAL-H to afford the corresponding aldehyde (**88**). Reaction with the anion from *C*-trimethylsilyl diazomethane gives the acetylene (**89**) in a single step. The chain is then extended by reaction of the acetylide anion with the triflate derivative from 3,3-dimethylbutanol. Exposure to strong acid serves to remove the triphenylmethyl protecting group on nitrogen. This last step affords **92**.[15] The absolute stereochemistry was derived from X-ray structure determination of one isomer of the sultam (**87**).

The three part so-called cocktail used to treat HIV positive patients typically comprise a proteinase inhibitor, such as those discussed in Chapter 1; a nucleoside-based reverse transcriptase inhibitor, such as those in Chapter 6, and a non-nucleoside inhibitor of reverse transcriptase (NNRTI). Most of the compounds in the first two classes share a good many structural features with other agents in the class. Chemical structures of the various NNRTIs on the other hand have little in common. **Capravirine (103)**, is notable in the fact that it fails to include any of the fused ring systems that provide the nucleus for other compounds in this class. Chlorination of 3-methylbutyraldehyde **(94)** provides one of the components for building the imidazole ring. For bookkeeping purposes, the condensation of **94** with *O*-benzyl glyoxal and ammonia can be envisaged as proceeding thorough the aminal (**95**) of the glyoxal. Imine formation with dichloro reagent **94** by displacement of halogen then leads to imidazole **96**. Reaction of that intermediate with iodine in base leads to the iodo derivative (**97**). Displacement of iodine by the anion from dichlorol sulfide (**98**) proceeds to give the thioether (**99**). The still-free imidazole nitrogen is next alkylated with 2-chloromethyl pyridine to afford **101**. The benzyl protecting group on oxygen is then removed by treatment with strong acid. The thus-revealed carbinol in **102** is condensed with chlorosulfonyl isocyanate to form the corresponding carbamate. Thus, the NNRTI **103** is obtained.[16]

The imidazole ring takes its place in this example among a wide variety of heterocyclic rings that serve as the nucleus for COX-2 NSAID anti-inflammatory compounds. Reaction of sulfonyl chloride (**104**), available from chlorosulfonation of acetanilide with *tert*-butylamine, gives the corresponding sulfonamide (**105**). The acetyl group on nitrogen is then removed by heating with strong base to give the aniline (**106**). Reaction of **106** with the fluoro anisaldehyde (**107**) gives imine **108**, which incorporates the two adjacent aromatic rings characteristic of COX-2 inhibitors. Reaction of the imine with toluenesulfonyl isocyanate in the presence of potassium carbonate leads to what may be viewed as 2 + 3 cycloaddition of the nitrogen analogue of a ketene to form the imidazole ring (**109**). This ring is then chlorinated with *N*-chlorosuccinimide (NCS) possibly to adjust the electron density on the heterocyclic ring. Heating this last intermediate (**110**) with acid removes the protecting group to give the free sulfonamide and thus **cimicoxib** (**111**).[17]

The enzyme dopamine-β-hydroxylase, as the name indicates, catalyzes hydroxylation of the side chain of dopamine in sympathetic nerves to form epinephrine. Direct antagonism of the enzyme shuts down production of that neurotransmitter. This achieves an effect on the cardiovascular system more directly than do either α- or β-blockers. The most immediate effect is manifested as a decrease in blood pressure. The synthesis of the specific hydroxylase inhibitor **nepicastat** (**122**) starts by reaction of aspartic acid with trifluoroacetic anhydride. This reagent results in conversion of the amine to its trifloroacetamide derivative and the acid to an anhydride (**113**). Reaction of this intermediate with 1,3-difluorobenzene in the presence of aluminum chloride gives the Friedel–Crafts acylation product (**114**). Catalytic hydrogenation then reduces the ketone to a methylene group (**115**). A second acylation reaction, this time via the acid chloride leads to the tertralone (**116**). The new carbonyl group is again reduced by means of hydrogenation; saponification then removes the protecting trifluoroacyl group to give the primary amine (**117**) as a single enantiomer. Reaction of that amine with formaldehyde and potassium cyanide leads to formation of what is essentially an α-aminonitrile, the nitrogen analogue of a cyanohydrin. The amino group is then taken to a formamide by reaction with butyl fomate. Formylation of the carbon adjacent to the nitrile by means of ethyl formate and sodium ethoxide puts into place the last carbon for the imidazole ring (**120**). Reaction of this last compound as its enolate with thiocyanate forms the cyclic thiourea (**121**). Catalytic hydrogenation serves to reduce the nitrile to the corresponding amino-methylene derivative and thus **122**.[18]

Collagenase enzymes are intimately involved in the destruction of cartilage that accompanies rheumatoid arthritis. Considerable attention has as a consequence been focused on finding inhibitors of that enzyme. The first step in the convergent synthesis starts by protection of the chiral hydroxy acid (**123**) as its benzyl ester (**124**). The hydroxyl is activated toward displacement by conversion to its triflate **125**. Reaction of **125** with the anion from the unsymmetrical malonate leads to triester **126**.

The α-aminonitrile (**127**) from acetone and methylamine comprises starting material for the heterocyclic moiety. Reaction of **127** with chlorosulfonyl isocyanate and hydrochloric acid gives hydantoin (**128**). Treatment of intermediate **128** with formaldehyde leads to a carbinol from addition to the free amino group on the imidazole dione. The hydroxyl group is then converted to the bromo derivative (**129**) with phosphorus tribromide.[19] Use of this intermediate (**129**) to alkylate the enolate from **126** yields **130**. Catalytic hydrogenation of this product leads to the formation of the corresponding ester-diacid by loss the benzyl protection groups on two of the esters. Heating this last intermediate in the presence of N-methylmorpholine causes the free acid on the carbon bearing the *tert*-butyl ester to decarboxylate (**131**). The desired stereoisomer (**131**) predominates, in effect reflecting the selectivity of alkylation step (**126** → **130**) caused by the presence of the preexisting adjacent chiral center. The free car-boxylic acid is condensed with piperidine to form **132**. The remaining ester is then hydrolyzed in acid to afford the acid (**133**). Reaction of **133** with *O*-benzylhydroxylamine followed by hydrogenolysis of the benzyl group then leads to the hydroxamic acid. Thus, the collagenase inhibitor **cipemastat** (**134**) is obtained.[20]

The discovery of canabinol receptors has led to the search for synthetic agonists and antagonists based on different structures from the hemp-related product. One of the first antagonists to come out of those programs, **rimonabant** (**140**) has shown activity as an appetite suppressant weight loss agent. Addition of the anion from the propiophenone (**135**) to the anion from ethyl oxalate gives the enolate (**136**). Condensation of that with 2,4-dichlorophenylhydrazine (**147**) results in formation of imines between carbonyl groups and the basic nitrogen thus forming the pyrrazole ring (**138**). Saponification of the ester affords the corresponding acid (**139**). This is then reacted with *N*-aminopiperidine in the presence of DCC to form the amide **140**.[21]

C. Thiazoles

A relatively simple thiazole has been shown to be a quite potent antiinfla-matory agent. **Darbufelone** (**143**), which is quite different in structure from all preceding NSAIDs inhibits both arms of the arachidonic acid cascade at the very inception of the process. This in effect shuts off production of both prostaglandins and leukotrienes. This agent is prepared in a single step by condensation of substituted benzaldehyde **141** with the enolate from thiazolone (**142**).[22]

141   142   143

Uric acid comprises one of the principal products from metabolism of endogenous nitrogen-containing compounds. Metabolic disorders that cause this pyrimidine base to accumulate in the bloodstream can cause gout, a painful condition that results from deposits of uric acid in joints. The uricosoric thiazole **febuxostat (147)**, like its venerable predecessor allopurinol, inhibits the enzyme xanthine oxidase, which is central to the production of uric acid. Febuxostat, whose structure is significantly different from its predecessor, has recently been introduced as treatment for gout. Reaction of the dinitrile compound (**144**) with hydrogen sulfide in base proceeds to convert the one of the two cyanide groups to the corresponding thioamide (**145**). Regioselectivity is speculatively due to reaction at the less hindered of the two nitriles. Condensation of this with the diketo-ester (**146**) leads to formation of the thiazole ring. Saponification of the ester completes the preparation of **147**.[23]

144   145   146   147

Reactive oxygen species released by neutrophils may play a role in conditions, such as inflammatory bowel disease and chronic pulmonary obstructive disease. A thiazole that inhibits *in vitro* production of superoxide by human neutrophils is currently being investigated in the clinic. In a convergent scheme, bromination of acyl pyridine carboxylic acid

(**148**) affords the acyl bromide. The product is then converted to the ester (**149**), by treatment with methanol in the presence of acid. Catechol esters undergo ready electrophillic attack as a result of the high electron density in the ring. Thus, reaction of diethyl catechol with isothiocyanate in the presence of acid leads to ring substitution. The initially formed thiocyanate hydrolyzes to the observed thiourea (**150**) under reaction conditions. In a classic method for forming heterocyclic rings, reaction of bromoketone (**149**) with the thiourea (**150**) proceeds to the thiazole **151**. Saponification of the ester then affords **tetomilast (152).**[24]

Heterocyclic compounds bearing nitro groups were among some of the earliest antiparasitic agents. A nitrothiazole has recently been approved for treating diarrhea due to such infections. This rather venerable compound, **nitazoxanide (155)** is prepared in a single step by reaction of the acid chloride (**153**) from aspirin with the aminonitrothiazole (**154**).[25]

Many peptides contain reasonably reactive amines, as well as an occasional free guanidine function. By the same token, the sugars that make up polysaccharide can be viewed as acetals of aldehydes. These functions on endogenous peptides and saccharides do occasionally interact chemically to form cross-links. Accumulation of cross-linked proteins with age is believed to lead to stiffening of tissues. Some of these processes may go as far as to result in pathologic changes. A structurally very simple thiazolium salt has shown activity in reversing such changes by breaking cross-links. The compound is probably prepared by reaction of dimethylthiazole (**156**) with phenacyl chloride (**157**). The product **alagebrium chloride (158)** is said to show promise in treating effects traceable to loss of tissue elasticity.

The "glitazones" comprise a large series of antidiabetic compounds that were introduced about a decade ago. The original hypoglycemic drugs used for control of Type 2 diabetes were marked by the presence of a sulfonyurea pharmacophore. This function is replaced by a thiazolidinedione group in the more recent glitazones. The synthesis of a very recent drug candidate in this group begins with reduction of the carboxylic acid in the naphthol (**159**) with diborane. The resulting carbinol is oxidized back to an aldehyde (**160**) by means of manganese dioxide. Aldol-type condensation of **160** with the active methylene group in thiazolidinedione itself leads to the unsaturated intermediate **161**. Next, catalytic hydrogenation serves to reduce the double bond. The free phenol in the other ring is then alkylated with *o*-fluorobenzyl chloride. Thus, the hypoglycemic agent **netoglitazone** (**163**) is obtained.[26]

D. Triazoles

Antifungal activity is retained in compounds in the "conazole" series when an additional nitrogen atom is inserted into the all-important heterocyclic ring. The preparation of a triazole-based antifungal agent starts with the construction of the pyrimidine ring. Thus, condensation of β-ketoester (**164**) with formamidine leads to pyrimidine **165**. Treatment of intermediate **165** with phosphorus oxychloride leads to the corresponding chlorinated compound (**166**). The key intermediate **168** could be obtained, for example, by alkylation of 1,2,4-triazole with phenacyl chloride (**167**). Addition of the enolate from treatment of the pyrimidine (**166**) with strong base to addition to the carbonyl group in **168**. The resulting tertiary alcohol (**169**) is obtained as a mixture of diastereomers. The chlorine atom, having served its function, is now removed by catalytic hydrogenation. Separation of diastereomers followed by resolution of the desired enantiomer pair affords the antifungal agent **voriconazole** (**170**).[27]

A rather more complex antifungal compound incorporates in its structure both a triazole and a triazolone ring. The lengthy sequence begins with displacement of chlorine in **171** by acetate. Reaction of the product with the ylide from methyl triphenylphosphonium bromide affords the methylene derivative (**173**). The acetate group is next saponified to give the free alcohol. The double bond is then oxidized in the presence of L-ethyl tartrate to afford epoxide **174** as a single enantiomer. The first heterocyclic ring is now introduced by opening of the oxirane with 1,3,4-triazole proper to afford **175**. Reaction with methenesulfonyl chloride gives the corresponding mesylate (**176**). Treatment with base leads to formation of an alkoxide on the teriary carbinol; internal displacement forms a new oxirane. This ring is then opened by the anion from diethylmalonate. The alkoxide that is formed as the initial product displaces the ethoxide group on one of the esters to form a lactone (**178**). Reduction with borohydride takes both carbonyl groups to alcohols affording the diol (**179**). This last intermediate (**179**) is again treated with toluenesulfonyl chloride to afford the bis(tosylate). Treatment with base leads to formation of an alkoxide from the still free tertiary alcohol. This compound undergoes internal displacement of one of the tosylate groups to form a THF ring (**180**). The remaining tosylate function serves as a leaving group for the next reaction. This last intermediate (**180**) is reacted with the diaryl piperazine (**181a**) in the presence of base.

The thus formed phenoxide displaces the toluenesulfonate to form the extended coupling product (**181b**). The nitro group is reduced to the corresponding amine. That function is then reacted with phenoxycarbonyl chloride to give the phenoxy carbamate. Treatment with hydrazine displace the phenoxide yielding the semicarbazone (**182**). Ethyl orthoformate supplies the remaining carbon atom to form the triazolone (**183**). The last step in the sequence comprises alkylation of the heterocyclic ring with the chiral methoxymethyl protected 2,3-pentanediol 3-tosylate (**184**). Thus, the antifungal agent **posaconazole (185)** is obtained.[28]

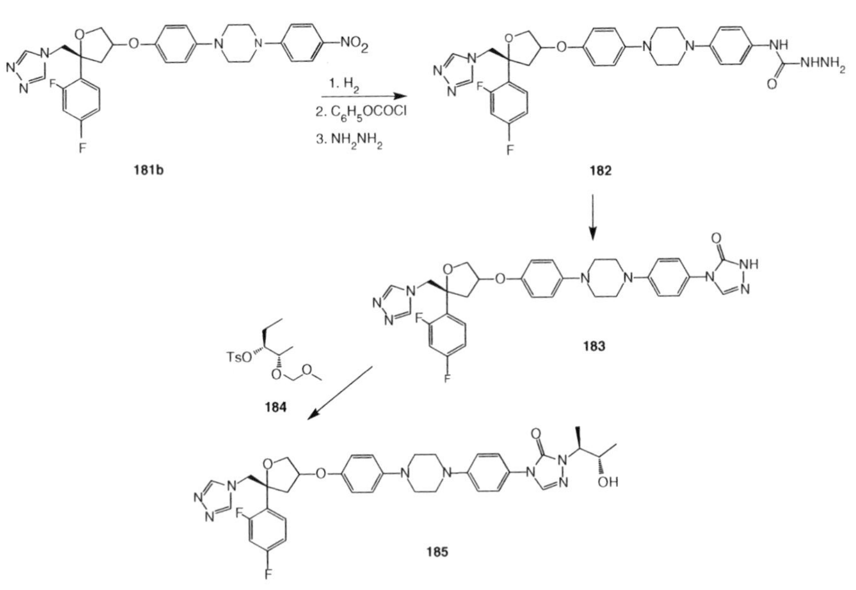

Administration of cancer chemotherapeutic agents is more often than not accompanied by serious bouts of nausea and vomiting. The serotonin antagonists, such as ondansetron, were the first class of antiemetic drugs to provide relief to patients undergoing chemotherapy . The involvement of substance P in mediation of the emesis reflex offers another target in the search for compounds for treating nausea. The demonstration that the substance P related neurokinin hNK-1 is directly involved in that reflex has led to the search for specific antagonists. The stereoselective synthesis of the antagonist **aprepitant (200)** begins with the preparation of chiral *p*-fluorophenylglycine (**190**). Coupling of the phenylacetic ester (**187**) with the chiral auxiliary (**186**) affords the amide (**188**). The requisite nitrogen atom is then introduced by treating the enolate from **188** with tosyl azide. Catalytic hydrogenation then reduces the azide to the corresponding primary amine (**189**). The chiral auxiliary, having done its work, is then removed. Thus hydrolysis with base leads to amino acid **190** as a single enantiomer. The benzyl protection group is next introduced by reductive alkylation with benzaldehyde. Reaction of the product (**191**) with 1,2-dibromoethane in the presence of mild base leads to formation of an ester with the carboxylic acid and then alkylation on nitrogen, though not necessarily in that order. The net result is formation of the morpholine ring (**192**). Treatment of the product with Selectride reduces the ester carbonyl to the aldehyde oxidation stage, present here as a cyclic acetal (**193**).

The ring hydroxyl group is then acylated with the fluorinated benzoyl chloride (**194**) to yield ester **195**. Reaction with the carbenoid species from the Tebbe reagent leads to formal replacement of carbonyl oxygen by a methylene group to form the enol ether (**196**). Catalytic hydrogenation reduces the enol to the corresponding ether at the same time deleting the benzyl protecting group. The presence of two adjacent chiral centers result in formation of product **197** as largely a single enantiomer. The remaining task involves formation of the pendant pyrazolone ring. Alkylation of the morpholine nitrogen with substituted semicarbazide **198** leads to **199**. Compound **199** undergoes internal displacement of the methoxy group on nitrogen, which results in formation of the triazolone ring and thus **200**.[29]

The virus that leads to AIDS is renowned for its ability to develop resistance to antiviral drugs. Successful treatment depends in some measure on finding agents that act on the virus by novel mechanisms. Current therapy thus combines drugs that act on three different stages of the viral

life cycle. A new class of agents depends on the fact that the virus needs to bind with specific receptor sites on the immune system cells in order to gain entry into these cells. The very recent antiviral agent **maraviroc** (**212**) binds to the same sites as HIV and thus prevents the very first stage in the process of infection. Synthesis of this agent starts by protection of the amino group in the bridged bicyclic amine (**201**). The carbonyl group at the other end of the molecule is then converted to its oxime. Treatment of this intermediate with sodium in alcohol reduces that group to a primary amine (**204**). Construction of the triazole ring first involves acylation of the amine with the acid chloride from isobutyric acid to form the isobutyramide (**205**). Reaction of **205** with phosphorus oxychloride converts the amide into the corresponding chlorinated imine (**206**). Treatment with acylhydrazide leads to addition–elimination of the basic hydrazide nitrogen to the imino chloride and thus formation of the imino–amide (**207**). Heating in the presence of acid leads to reaction of the imino nitrogen with the carbonyl group. This closes the ring and affords the triazole (**208**). Catalytic reduction removes the benzyl protecting group ummasking the basic ring nitrogen (**209**). In a converging scheme, the ester in the peptide-like fragment (**210**) is reduced to afford aldehyde **211**. Reductive amination of **211** with amine **209** and sodium triacetoxy borohydride leads to the coupled product **212**.[30]

A substituted 1,2,3-triazole ring provides the pharmacophore for an antiepileptic drug. Reaction of the 2,5-difluorobenzyl bromide (**213**) with sodium azide leads to displacement of the benzylic halogen and formation of azide **214**. Treatment of **214** with propargylic acid leads to a 2 + 3 cycloaddition reaction and thus formation of the 1,2,3-triazine ring (**215**). The carboxylic acid is then converted to its amide via the acid chloride. Thus, the antiepileptic agent **rufinamide** (**216**) is obtained.[31]

As noted earlier, protein kinases play a pivotal role in cell proliferation. Inhibitors of these enzymes show promise as antitumor agents particularly, those that show preference for malignant cells. The kinase inhibitor **mubritinib** (**221**) is currently being evaluated as a drug for treating breast cancer. The first step in the convergent synthesis comprises displacement of a leaving group, such as methanesulfonate, from a suitably protected phenol (**217**) by 1,2,3-triazine proper. Removal of the protecting group leads to the free phenol (**218**). Preparation of the second moiety involves reaction of one of the classical methods for forming of an oxazole: reaction of a halomethyl carbonyl group with an amide. Thus, condensation of the cinnamic amide (**219**) with 1,3-dichloroacetone leads to formation of oxazole **220**, which retains a leaving group for a displacement reaction. Treatment of **220** with the alkoxide from treatment of **218** with base leads to the corresponding ether (**221**).[32]

E. Tetrazoles

Though the exact cause of Alzheimer's disease is still unclear, evidence points to the utility of increasing acetylcholine (AcCh) levels for treating that condition. Most approaches are aimed at devising inhibitors of cholinesterase, the enzyme that destroys AcCh. A quite different tack involves developing compounds that have cholinergic activity in their own right. The tetrazole **alvameline** (**230**), for example, was developed as a bioisostere of the muscarinic cholinergic compound arecoline (**222**). The design devolves on the fact that the proton on a free tetrazole shows a pK_a comparable to that of a carboxylic acid. Fully substituted tetrazoles as in **230**, may thus in some ways may be viewed as surrogate esters. Alkylation of nicotinonitrile (**223**) with methyl iodide affords methiodide **224**. Treatment of this intermediate with borohydride reduces it to tertrahydropyridine (**225**) in which the position of the double bond mimics that in arecoline. Reaction of **225** with ethyl chloroformate results in N-demethylation and consequent formation of the corresponding carbamate. The nitrile group is then transformed to a tetrazole by reaction with sodium azide in the presence of aluminum chloride, one of the standard procedures for building that ring. The surrogate acid is then alkylated with ethyl iodide to afford **228**. Treatment with acid then removes the carbamate on the ring nitrogen (**229**). The methyl group on the piperidine ring is restored by reaction with formaldehyde and formic acid, under standard Clark–Eshweiler conditions. Thus, the muscarinic agonist **230** is obtained.[33]

A neurokinin inhibitor whose structure differs markedly from aprepitant (**200**) incorporates a substituted tetrazole ring. The synthesis of the tetrazole-containing moiety of **vofopitant** (**241**) start by acylation of substituted aniline **231** with trifluoroacetyl chloride to afford the amide (**232**). Reaction of that under Mitsonobu conditions leads to the enol chloride (**233**). Treatment of **233** with sodium azide probablty starts with addition–elimination of azide ion; this undergoes internal 1,3-cycloaddition to form the tetrazole ring. Catalytic hydrogenation then removes the benzyl protecting group to reveal the free phenol (**235**). Reaction of **235** with hexamethylene tetramine (HMTA) in acid leads to formylation at the ortho position to give the substituted hydroxybenzaldehyde (**236**). The phenol is next converted to the corresponding methyl ether (**237**), by alkylation with methyl iodide in the presence of base.

Construction of the second part of the molecule starts with palladium-catalyzed coupling of the substituted pyridine (**238**) with phenylboronic acid to give **239**. Hydrogenation reduces both the nitro group and the supporting pyridine ring to afford **240** as the cis isomer. The enantiomers of this are then separated by resolution. The desired isomer is then subjected to reductive amination with aldehyde **237** affording **241**.[34]

The renin-angiotensin system plays a profound role in maintaining the circulatory system. Elevated levels of angiotensin II are closely associated with hypertension. Angiotensin converting enzyme inhibitors, the so-called ACE inhibitors, lower blood pressure by preventing generation of that enzyme from its precursor. These now-widely used drugs were first introduced starting almost three decades ago. Attention has turned more recently to drugs that act directly on angiotensin receptors. The preparation of one of these non-peptide agents begins with the formation of an imidazole. Thus, condensation of the diamine **242** with trimethoxy butyl-orthoformate affords the heterocycle **243**. The nitrile groups are then hydrolyzed to the corresponding acids; these groups are then esterified with ethanol (**244**). Reaction of intermediate **244** with methylmagnesium bromide leads to addition of but one of the ester groups. The presence of the initial charged adduct arguably hinders addition to the second ester. The free amino group on the imidazole is then allowed to react with the benzylic halide on the tetrazole-substituted biphenyl (**246**) to afford the alkylated product (**247**). Saponfication of the ester and removal of the triphenylmethyl protecting group compeles the synthesis of the angiotensin antagonist **olmesartan** (**248**).[35]

REFERENCES

1. E.B. Villhauer, J.A. Brinkman, G.B. Nader, B.F. Burkey, B.E. Dunning, K. Prasad, B.L. Mangold, M.E. Russel, T.A. Hughes, *J. Med. Chem.* **46**, 2774 (2003).

2. S.J. Coutts et al., *J. Med. Chem.* **39**, 2087 (1996).

3. S.J. Wittenberger, M.A. McLaughlin. *Tetrahedron Lett.* **40**, 7175 (1999).

4. P.E. Cross, A.R. MacKenzie, U.S. Patent 5,096,890 (1992).

5. M.M. Faul, E. Kobierski, M.E. Kopach, *J. Org. Chem.* **68**, 5739 (2003).

6. W.F. Heath, J.H. McDonald, M. Paal, T. Schotten, W. Stenzel, U.S. Patent 5,672,618 (1997).

7. M.M. Faul, J.R. Gillig, M.R. Jirousek, L.M. Ballas, T. Schotten, A. Kahl, M. Mohr, *Bioorg. Med. Chem. Lett.* **13**, 1857 (2003).

8. C.D.W. Brooks et al., *J. Med. Chem.* **38**, 4768 (1995).

9. J. Haruta, H. Hashimoto, M. Matsushita, U.S. Patent 5,994,381 (1999).

10. J.J. Talley et al., *J. Med. Chem.* **43**, 775 (2000).

11. J.J. Talley et al., *J. Med. Chem.* **43**, 1661 (2000).

12. R.R. Bartlett, F.-J. Kammerer, U.S. Patent 5,949,511 (1996).

13. P.T. Gallagher, T.A. Hicks, G.W. Mullier, U.S. Patent 4,892,963 (1990).

14. K. Thiele, L. Zimgibl, M.H. Pfeninger, M. Egli, M. Dobler, *Helv. Chim. Acta* **70**, 441 (1987).

15. H. Liu, F.A. Kerdesky, L.A. Black, M. Fitzgerald, R. Henry, T.A. Ebenshade, A.A. Hancock, Y.L. Bennani, *J. Org. Chem.* **69**, 192 (2004).

16. H. Sugimoto, T. Fujiwara, U.S. Patent 6,147,097 (2000).

17. C. Almansa et al., *J. Med. Chem.* **46**, 3463 (2003).

18. G.R. Martinez, O.W. Goodling, D.B. Repke, P.J. Teitelbaum, K.A.M. Walker, R.L. Whiting, U.S. Patent 5,719,280 (1998).

19. H.R. Wiltshire, K.J. Prior, J. Dhesi, G. Maile, *J. Labelled Cpd. Radiopharm.* **44**, 149 (2001).

20. M.J. Broadhurst et al., *Bioorg. Med. Chem. Lett.* **7**, 2299 (1997).

21. F. Barth, P. Casellas, C. Congy, S. Martinez, M. Rinaldi, G. Anne-Archard, U.S. Patent 5,624,940 (1997).

22. W.A. Cetenko, D.T. Connor, J.C. Sircar, R.J. Sorenson, P.C. Unangst, U.S. Patent 5,143,928 (1992).

23. T.A. Robbins, H. Zhu, J. Shan, U.S. Patent Appl. 2005/0075503 (2005).

24. M. Chihiro et al., *J. Med. Chem.* **38**, 353 (1995).

25. J-.F. Rossignol, R. Cavier, U.S. Patent 3,950,351 (1976).

26. L.A. Sobrera, J. Castaner, M. del Fresno, J. Silvestre, *Drugs Future* **27**, 132 (2000).

27. R.O. Fromtling, *Drugs Future* **21**, 266 (1996).

28. J. Heeres, L. Backx, J. Van Custem, *J. Med. Chem.* **27**, 894 (1984).

29. J.J. Hale et al., *J. Med. Chem.* **41**, 4607 (1998).

30. M. Perros, D.A. Price, B.L.C. Stammen, A. Wood, U.S. Patent 6,667,314 (2003).

31. L.A. Sobrera, P.A. Leeson, J. Castaner, *Drugs Future* **25**, 1145 (1998).

32. A. Tasaka, T. Hitaka, E. Matsutani, U.S. Patent 6,716,863 (2004).

33. E.K. Moltzen, H. Pedersen, K.P. Bogeso, E. Meier, K. Frederiksen, C. Sanchez, H.L. Lambol, *J. Med. Chem.* **37**, 4085 (1994).

34. D.R. Armour et al., *Bioorg. Med. Chem. Lett.* **6**, 1015 (1996).

35. H. Yanagisawa et al., *J. Med. Chem.* **39**, 323 (1996).

CHAPTER 6

SIX-MEMBERED HETEROCYCLES

1. COMPOUNDS WITH ONE HETEROATOM

A. Pyridines

Agents that interact with cholinergic receptors have been extensively investigated for the treatment of Alzheimer's disease. Most of the compounds discussed to this point bind at muscarinic sites. An alternate approach involves administration of agents that bind to cholinergic nicotinic receptors. A compound closely related structurally to nicotine itself is currently being investigated in the clinic. The synthesis of this agent starts with the dihalogenated pyridine (2) obtained from nicotine (1). Coupling of this with the monosilyl derivative from acetylene in the presence of palladium: triphenyl phosphine and copper iodide replaces iodine on the aromatic ring by the acetylene moiety 3. The chlorine on the ring is then removed by reduction with zinc in acetic acid 4. The silyl protecting group is then cleaved with fluoride ion to afford **altinicline (5)**.[1]

The Organic Chemistry of Drug Synthesis, Volume 7. By Daniel Lednicer
Copyright © 2008 John Wiley & Sons, Inc.

As noted several times in Chapter 5, a five-membered ring containing one or more heteroatoms comprises the central element in a great majority of COX-2 inhibitor NSAIDs. Interestingly, that moiety can be replaced by a six-membered pyridine ring. One of the two adjacent rings that characterize these agents in this case include a nitrogen atom. One of the routes to this agent starts with the nicotinic ester (**6**). The ester group is then reduced by means of DIBAL-H; back oxidation with manganese dioxide affords aldehyde **7**. Reaction with aniline and diphenylphosphate gives adduct **8** in a reaction that parallels formation of a cyanohydrin. Treatment of intermediate **8** with strong base generates an ylide on the carbon bearing nitrogen. Condensation of **8** with benzaldehyde **9** leads to enamine **10** which now contains two of the requisite rings. Hydrolysis of **10** leads to the ketone (**11**). The central ring is formed by a newly devised pyridine synthesis. The first step involves reaction of chloroacetic acid (**12**) with dimethylformamide (DMF); the initial adduct is then converted to its hexafluorophosphate salt (**13**). Condensation of **13** with ketone **11** involves initial displacement of one of the dimethylamino groups by the enolate from **11** to give the open-chain adduct (**14**). Reaction with ammonium hydroxide then closes the ring forming the required pyridine. Thus, the NSAID **etoricoxib** (**15**) is obtained.[2]

B. Reduced Pyridines

Much effort has been devoted over the years to devise long acting forms of drugs. The occasion does, however, sometimes arise where blood levels of a drug being administered by infusion need to be cut off abruptly. Several examples, such as the β-blocker esmolol or the analgesic agent remifentanyl, include in their structure a function that is quickly converted to a carboxylic acid. This functional group both inactivates the drug and hastens its excretion. Dihydropyridines comprise a large group of calcium channel blockers used to treat angina and hypertension. The short-acting example, **clevidipine** (**18**) incorporates a methoxymethyl ester, a function that is readily cleaved to the corresponding acid by serum esterases. The actual experimental details in patents[3,4] describe the preparation of this compound by simple alkylation of the monoester (**16**) with chloromethyl butyrate (**17**). These sources are, however, mute as to source of the half ester starting material.

A very similar strategy underlies the design of the opiate-like analgesic **alvimopan** (**30**). Ileus, that is paralysis of the gastrointestinal tract, is a common side effect from surgery. The shock of the operation accompanied by heavy use of opiates causes a shut down of intestinal peristalsis. The structure of that drug in effect comprises a very potent synthetic opiate modified by a polar glycine residue. That last moiety keeps the drug out of the CNS. Reversing postoperative ileus comprises one of the main applications of this compound. The synthesis of this compound begins with addition of the Grignard reagent from substituted the bromobenzene (**20**) to the piperidone (**19**). The hydroxyl group is then acylated with ethyl chloroformate. Heating the resulting ester (**22**) leads to formation of the styrene (**23**). Treatment with base under equilibrium conditions leads to migration of the negative charge to the quarternary carbon adjacent to the aromatic ring. Addition of dimethyl sulfate thus leads to alkylation at that position.[5] Reaction of **24** with sodium borohydride leads to reduction of what is now an enamine, and thus formation of the saturated piperidine (**25**). The methyl group is then removed using chloroformate in the modern version of the Von Braun reaction. Treatment of product **26**, with methyl acrylate leads to Michael addition and formation of **27**. The carbon adjacent to the ester is next converted to its enolate with lithium diisopropylamine (LDA); addition of benzyl bromide leads to alkylation and formation of **28**. The ester (**29**) is then saponified. Condensation with glycine ester followed by saponification yields **30**.[6]

Multidrug resistance, where a tumor becomes immune to a broad range of compounds without regard to their mechanism of action, comprises one of the more important pitfalls in cancer chemotherapy. Considerable effort has been devoted to finding drugs, so-called chemosensitizing agents, that reverse the process. The synthesis of one of these, **biricodar** (**38**) is presented in outline form omitting several protection–deprotection steps. One arm of the convergent scheme begins with the oxidation of trimethoxyacetophenone (**31**) with selenium dioxide to afford the corresponding acid (**32**). Compound **32** is then condensed with the silyl ester of pipecolic acid to afford the amide; deprotection yields acid **33**. Construction of the second moiety involves first addition of the addition of an organometallic derivative of propargyl bromide to but-3-ynal (**34**). The anion from removal of the acetylenic protons with strong base is then treated with 3-bromopyridine. This affords the product (**36**) from displacement of halogen by the terminal acetylenic carbon. Catalytic hydrogenation then leads to intermediate **37**. Esterification of the acid (**33**) with alcohol **37** affords (**38**).[7]

C. Miscellaneous

The design of the peptidomimetic antiviral agent (Chapter 1) ultimately traces back to the fact that structures of these protease inhibitors in some way act as surrogates for the natural substrate enzyme. The recent protease inhibitor antiviral drug **tiprinavir** (**52**) notably departs from this pattern as it omits the peptide-like moiety present in the earlier agents; this drug is also as a result said to be active against HIV strains that have developed resistance to the peptidomimetic agents. The convergent synthesis begins with the addition of the enolate from methyl acetate to the carbonyl group in **39**. The product (**40**) is then saponified and the resulting acid resolved by way of its ephedrine salt. Treatment of the desired enantiomer with chloromethyl *p*-hydroxybiphenyl gives the doubly alkylated derivative **41**. Reaction of that compound with DIBAL-H results in reduction of the ester to an alcohol (**42**). This function is then back-oxidized with hypochlorite in the presence of tetramethylpiperidol *N*-oxide to give the aldehyde (**43**).

Preparation of the other major fragment begins with Knoevenagel condensation of *m*-nitrobenzaldehyde with dimethyl malonate to yield diester **45**. Treatment of intermediate bromide **45** with diethyl zinc in the presence of cupric iodide leads to conjugate addition of an ethyl group (**46**). The diacid obtained on saponification of that product then spontaneously decarboxylates on warming. The product obtained on esterifying the remaining carboxylic function acid is next resolved by chromatography over a chiral column. Condensation of the enolate from treatment of **47** with the aldehyde in the other fragment affords the adduct (**48**). The newly introduced hydroxyl group is oxidized to the ketone by means of pyridinium chlorochromate to afford the β-ketoester (**49**). The biphenoxy protecting group on the tertiary alcohol is then removed by acid hydrolysis.

Reaction of this intermediate with base leads the anion of the newly revealed hydroxyl group to attack with the carboxylate in what amounts to an internal transesterification. This step forms the cyclic ester and thus the requisite pyranone ring (**50**). Catalytic hydrogenation of **50** results in reduction of the nitro group to yield **51**. Acylation of the newly formed amine with 5-trifluoromethylpyridinium-2-sulfonyl chloride affords **52**.[8]

2. COMPOUNDS WITH TWO HETEROATOMS

A. Pyrimidines

The structures of all currently approved gastric acid secretion inhibitors that act as inhibitors of the sodium–potassium pump consists of variously substituted pyridylsulfonyl-benzimidazoles. A structurally very distinct compound based on a pyrimidine moiety has much the same activity as the benzimidazole-based drugs. In yet another convergent synthesis, reaction of β-phenethylamine (**53**) with acetic anhydride affords amide **54**. Treatment with polyphosphoric acid (PPA) then leads to ring closure to form the dihydroisoquinoline (**55**). Sodium borohydride then reduces the enamine function to afford fragment **56**.

In a classic sequence for building a pyrimidine ring, the aniline (**57**) is condensed with cyanamide. Addition of the basic nitrogen to the nitrile in the reagent leads to formation of the guanidine (**58**). Condensation of **58** with methyl ethyl acetoacetate results in formation of the pyrimidine (**59**); the residue of the carboxyl group appears as an enol oxygen (**59**; R = OH). Treatment of this intermediate with phosphorus oxychloride replaces the enol oxygen by chlorine. Displacement of this last group by the basic amine in other moiety (**56**) leads to the coupling product (**61**). Thus, the sodium–potassium pump inhibitor **revaprazan** is obtained.[9]

A pyrimidine ring forms the nucleus for yet another nonnucleoside reverse transcriptase inhibitor (NNRTI) active against HIV. This heterocyclic ring is prepared in a manner analogous to that outlined above. The starting guanidine (**62**) can be prepared by reaction of 4-cyanoaniline with cyanamide. Condensation of **62** with ethyl malonate leads to the substituted pyrimidine **63** in a single step. The enolic hydroxyls are then replaced by chlorine by reaction with phosphorus oxychloride. Treatment of product **64** with bromine replaces the remaining hydrogen on the pyrimidine ring by bromine (**65**). Reaction of intermediate **65** with the enolate from 2,6-dimethyl-4-cyanophenol displaces one of the halogens forming an ether linkage (**66**). The symmetrical nature of the pyrimidine ring renders moot regiochemistry. Displacement of the remaining chlorine with ammonia completes the synthesis of **etravirine** (**67**).[10]

The structure of the relatively simple pyrimidone **emivirine** (**71**) might suggest that this compound is a classical non nucleoside reverse transcriptase inhibitor. Detailed studies have, however, shown that instead this compound acts at the same site as other NNRTIs. Base-catalyzed alkylation of the pyrimidine (also called a uracil) (**68**) with ethyl chloromethyl ether perhaps surprisingly takes place at the nitrogen flanked by a single carbonyl to yield **69**. Reaction with a second equivalent of base, this time LDA, followed by benzaldehyde results in addition of a benzyl group to the only open-ring position. The reaction is then quenched with acetic anhydride to afford the acetoxy derivative (**70**). Catalytic hydrogenation then removes the acetoxy group to afford **71**.[11]

The efficacy and good tolerance of so-called statins has led to their widespread use in the treatment of elevated serum cholesterol. These agents act as a very early step in the endogenous synthesis of that steroid. This initial stage comprises reduction of the activated carboxylic acid group (COSCoA) (CoA is coenzyme A) in the glutaric acid derivative hydroxymethylglutaryl CoA to an alcohol (CH_2OH); the product, mevalonic acid, then goes on to the isoprene equivalent, isopentenyl pyrophosphate, in several more steps. This pivotal reaction is catalyzed by an enzyme hydroxymethylglutaryl CoA reductase (HMG–CoA); statins are thus more correctly classed as HMG–CoA inhibitors. All approved inhibitors feature a side chain that mimics the HMG–CoA substrate. The remainder of the structure of drugs varies widely ranging from the decalins found in the first statins produced by fermentation to monocyclic heterocyclic rings in some more of the recently introduced drugs. The synthesis of a statin based on a pyrimidine ring begins with Knoevnagel condensation of methyl 4-methylacetoacetate (**73**) with *p*-fluorobenzaldehyde to afford the acrylate (**74**). This last intermediate is then reacted with thiourea. In a variation on the standard scheme for preparing pyrimidines. The reaction in this case involves conjugate addition to the double bond followed imine formation with the ketone though not necessarily in that order. The resulting dihydropyrimidine is then dehydrogenated with dichloro-dicyanoquinone (DDQ) to afford **75**. The thioether function is next oxidized to the sulfoxide by mean of *m*-perbenzoic acid (MCPBA) to afford **76**. Treatment with methylamine replaces the sulfoxide by nitrogen by an addition–elimination sequence. The thus-introduced amino group is then converted to the sulfonamide (**77**) by reaction with methanesulfonyl chloride. The ester group is then reduced to the corresponding alcohol by means of DIBAL-H and back-oxidized to aldehyde (**78** = **79**) required for attachment of the side chain.

Condensation of aldehyde **79** with the ylide from chiral phosphorane (**80**) attaches the moiety that will become the statin side chain in a one fell swoop (**81**). The silyl protecting group on the alcohol is removed next by treatment with hydrogen fluoride. The side-chain ketone is reduced using a mixture of sodium borohydride and diethyl methoxyboron. This last reagent forms a chelate between the ketone and hydroxyl group in the starting material, in effect transferring the chirality of the hydroxyl to that which is to be introduced. The diol product (**82**) thus comprises a single enantiomer. Saponification of the ester group affords **rosuvastatin** (**83**) as its sodium salt.[12]

Variously substituted benzyl diamino pyrimidines have a venerable place in the history of antibiotics. These drugs, exemplified by the still widely used, trimethoprim, are known to inhibit the enzyme dihydrofolate reductase, crucial for bacterial replication. A very potent recently introduced example has shown promise for treating infections by antibiotic resistant bacteria. The synthesis of a fragment necessary for building a fused ring begins with aluminium chloride catalyzed acylation of bis-sylilated acetylene (**84**) with carboxycyclopropyl chloride. The reaction interestingly stops with the introduction of a single acyl function (**85**). The newly introduced ketone is then reduced to the corresponding alcohol (**86**), with sodium borohydride. Mitsonobu reaction of this alcohol with phenol **87** leads to the alkylation product, ether (**88**). This product undergoes a internal $2 + 4$ electrocyclic addition on heating forming a dihydropyran ring and thus product **89**. The ester grouping is then subjected to a reduction-back oxidation sequence in order to convert ester group to the corresponding aldehyde (**90**). Aldol condensation of the 3-anilidopropionitrile carbonyl function, probably proceeds initially to adduct **91**. The double bond then shifts out of conjugation with the aromatic ring to give **92** as the observed product. The reaction of intermediate **92** with guanidine can be viewed for bookkeeping purposes as involving addition of one of the guanidine nitrogens to the nitrile while another displaces the anilide by an addition–elimination step to close the pyrimidine ring. Thus, the antibacterial agent **iclaprim** (**93**) is obtained.[13]

The search for endothelin antagonists as potential compounds for treating cardiovascular disease was noted in Chapter 5 (*see* atrasentan). A composed with a considerably simpler structure incorporates a pyrimidine ring in the side chain. Condensation of benzophenone (**94**) with ethyl chloroacetate and sodium methoxide initially proceeds to addition of the enolate from the acetate to the benzophenone carbonyl. The alkoxide anion on the first-formed quaternary carbon then displaces chlorine on the acetate to leave behind the oxirane in the observed product (**95**). Methanolysis of the epoxide in the product in the presence of boron trifloride leads to the ether–alcohol (**96**). Reaction of this with the pyrimidine (**97**) in the presence of base leads to displacement of the methanesulfonyl group by the alkoxide from **96**. Saponification of the ester group in that product gives the corresponding acid, **ambrisentan (98)**.[14]

An entire new field of therapeutics arose with the serendipitous discovery of the effect of sildefanil, far better known as Viagra, on erectile function. The almost predictable wild success of that drug has led to the search for other inhibitors of phosphodiesterase 5, the enzyme responsible for this activity. The structures of many follow-on agents hewed fairly close to the original PDE 5 inhibitor. Others, such as **avanafil (107)**, differ markedly in structure from sildefanil. The synthesis of agent **107** in effect consist of a series of displacement reactions. Thus, reaction of the benzylamine (**99**) with chloropyrimidine (**100**) leads to displacement of chlorine and formation of the coupling product (**101**). The inert methylthio ether on the pyrimidine ring is made labile by conversion to a sulfoxide (**102**) by means of *m*-chloroperbenzoic (*M*CPBA) acid. Reaction of that product with L-prolinol (**103**) leads to the replacement of the sulfoxide by the basic nitrogen in **103** and formation of coupling product **104**. The ester group in this last intermediate is then hydrolyzed to afford the corresponding acid (**105**). Reaction of this acid with 1-aminomethylpyrimidine (**106**) in the presence of a carbodiimide leads to the amide (**107**).[15]

The duration of action of venerable antitumor antimetabolite 5-fluoruracyl (5-FU, **108**) is limited by its relatively fast destruction by liver enzymes. One approach to extending the half-life of the drug involves a prodrug in which the sites where degradation occurs are covered by moieties known to inhibit the metabolism of other uracils. The majority of the drug in circulation would then consist of the prodrug. Slow loss of these protecting groups would then lead to extended therapeutic levels of active 5-FU. The first few steps in preparing **emitefur (116)** involves a scheme analogous to that used to prepare the prodrug tegafur. In this sequence, 5-FU is first converted to the silylated derivative (**109**) by means of trimethylsilyl chloride. Treatment of that with ethyl chloromethyl ether in the presence of stannic chloride leads to ethoxymethylated intermediate **110** with loss of the silyl groups. This compound (**110**) is then acylated with the bis acid chloride (**111**). In a convergent step, the pyridinol (**113**) is acylated with benzoyl chloride; reaction takes place at the sterically more open hydroxyl to afford monobenzoate (**114**). This compound is then allowed to react with the acid chloride (**112**) under more forcing conditions. Acylation of the remaining hydroxyl then affords the prodrug **115**.[16]

An important stage in the process of cell division comprises the formation of structures called microtubules that will pull apart the halves of the dividing cell nucleus in the course of replication. Microtubules normally dissolve after they have fulfilled their function. The very effective chemotherapy drug paclitaxel in effect stabilizes microtubules and halts cell division in mid-process. The structurally relatively simple compound **monasterol** (**120**) hinders formation of tubules at their very inception by inhibiting a required enzyme. This compound is prepared in a single step by multicomponent reaction of *m*-hydroxybenzaldehyde (**117**), ethyl acetoacetate (**118**), and thiourea (**119**).[17] One way this transform can be visualized is to assume that the acetoacetate enolate adds to the aldehyde. The second step involves conjugate addition of one of the urea nitrogen to the resulting enone; reaction of the other nitrogen with the ketone group would then close the ring.

Half of the bases in RNA and DNA consist of pyrimidine nucleosides. Analogues based on those nucleosides have, as result, been intensively investigated as sources for both antitumor and antiviral compounds. Such analogues, whether modified on the sugar portion or the base itself could in theory act as false substrates for processes dependent on those nucleotides at target cells, and lead to their demise. Replacement of one of the hydrogen atoms in the sugar portion of cytidine by fluorine results in a compound that has shown activity against the hepatitis B virus.

The preparation of this agent starts with the fully protected arabinose derivative (**121**). Treatment of **121** with hydrogen bromide leads to migration of the benzoyl group to the anomeric position adjacent to the ring oxygen with displacement of the acetate (**122**). The free ring hydroxyl function is then converted to a good leaving group by reaction with sulfuryl chloride followed by pyrrole to yield **123**. Reaction of product **123** with tetrabutylammonium fluoride leads to displacement of the sulfonamide fragment by fluorine with inversion of configuration. The anomeric acetate (**124**) is next replaced by bromine by means of hydrogen bromide. Compound **125** is then allowed to react with silylated thymine (**126**); this leads to the glycosylated base **127** with loss of the silyl groups. Treatment of intermediate **127** with mild base leads to hydrolysis of the last benzoyl protecting groups. Thus, the antiviral agent **clevudine** (**128**) is obtained.[18]

A cytosine derivative with a highly modified sugar showed significant antitumor activity in a variety of model systems. This activity did not, however, hold up in the clinic resulting in discontinuation of further clinical trials.[19] Reaction of cytidine (**129**) with the bis(chlorotetraisopropylsiloxane) (**130**) results in reaction with the hydroxyl groups at positions 3 and 5 of the sugar, forming a bridged structure in which the oxygen atoms at those positions are protected. The difunctional reagent is apparently too long to form the more common 2,3 cyclic protected derivative. Oxidation by means of oxalyl chloride in DMF then gives ketone **132**. That function is then condensed with the ylide from the complex phosphonate **133** to afford. The substituted exomethylene derivative (**134**). Reaction of **134** with tributyl tin hydride results in replacement of the phenyl sulfone by tributyl tin (**135**). Treatment of **135** with potassium fluoride results in cleavage of the silicon–oxygen bonds, as well as replacement of tin by hydrogen. This reaction leads to the antimetabolite **tezacitabine** (**136**).[20]

Zidovudine, better known as AZT, was the first drug that showed any activity against HIV. This compound comprises thymidine in which an azide group replaces hydroxyl on the sugar moiety. The finding that this agent was effective against HIV set off the search for additional reverse transcriptase inhibitors in the hope that they would be active against strains of the virus that had developed resistance to AZT. Replacement of one of the carbon atoms in the saccharide by another element has proven a particularly fruitful stratagem. Plentiful ascorbic acid (**137**) comprises the starting material for an early synthesis for **troxacitabine** (**146**). Reaction of **137** with the acetal (**138**) forms the cyclic acetal (**139**) from the two terminal hydroxyls as a mixture of enantiomers. Oxidation with peroxide cleaves the furenone ring with loss of a carbon atom to afford the hydroxy acid (**140**). Ruthenium trichloride next cleaves off the terminal carboxyl group to give **141**. The two isomers are then separated by flash chromatography. Oxidation with lead tetraacetate disposes of the final carboxyl group with concomitant transfer of an acetate (**142**). Catalytic hydrogenation results in loss of the benzyl protecting group. The thus-revealed hydroxyl group is acylated to afford the diacetate (**143**). Glycosylation with cytidine involves the by now familiar theme. Thus, reaction of acetate

142 with the silylated derivative 144 in the presence of Lewis acid leads to the coupling product 145 with loss of the silyl groups. Separation of the diastereomers followed by treatment with dilute base affords the reverse transcriptase inhibitor (146).[21]

Replacement of one of the sugar carbon atoms by sulfur leads to the HIV reverse transcriptase inhibitor **emtricitabine (151)**. As an additional departure from the natural nucleoside, this agent features a fluorine atom on the pyrimidine ring. This compound (151) can be prepared by a scheme analogous to that used to prepare the des-fluoro predecessor lamivudine. Thus, reaction of the benzoyl ester of glycolaldehyde (147) with the dimethyl actetal thioglycolaldehyde (148), affords the surrogate sugar (149) in a single step. This condensation may be viewed as starting with addition of sulfur to the free carbonyl in aldehyde 148. Displacement of one of the methoxy acetals by the newly formed hydroxyl closes the ring. That product is next reacted with the trimethylsylilated cytidine derivative 150 in the presence of a Lewis acid to afford the glycosilated fluorocytosine 151. Separation of diastereomers followed by removal of the benzoate protecting group affords 151.[22]

Migraine headaches are quite resistant to conventional peripheral or central analgesics. The discovery of the indole-based triptamine HT_{1D} antagonists, such as sumatriptan for the first time provided a means for relieving attacks of this sometimes disabling condition. The discovery of a structurally unrelated antagonist broadened the SAR for the design of triptamine HT_{1D} antagonist. Reaction of the benzopyran acid (152) with thionyl chloride leads to the corresponding acid chloride. Hydrogenation of a solution of that intermediate in thiophene stops at the aldehyde stage (153). Reductive amination with benzylamine gives the secondary amine (154). Treatment with acrylonitrile leads to conjugate addition and formation of 155. Exhaustive hydrogenation of that product leads to reduction of the nitrile to a primary amine and at the same time removes the benzyl protecting group. Thus, the diamine (156) is obtained. In one scheme for adding the terminal heterocycle, the diamine is alkylated with 2-chloropyrimidine (**157**) to afford **158**. Yet another hydrogenation reduces the heterocyclic ring to a tetrahydropyrimidine (**159**). The three terminal nitrogen atoms in the product **alniditan** (**159**)[23] in effect comprise a cyclized guanidine.

B. Miscellaneous Six-Membered Heterocycles

The proteasome is an enzyme complex found in all cells that is responsible for the turnover of the proteins regulating the cell cycle, as well as other cellular processes. This complex becomes unregulated in tumor cells and leads to excessive degradation of cycle regulatory proteins and tumor suppression genes. The absence of these factors leads to run-away cell proliferation. An unusual short peptide-like compound in which a boronic acid replaces the usual terminal carboxylic acid has proven to be an effective blocker of the proteasome. The first step in preparing that compound comprises amide formation between the aminoboronic acid (**161**), where the boron is protected as the cyclic pinanediol ester, and the *tert*-butyloxycarbonyl (*t*-BOC) protected phenylalanine (**160**) using standard peptide chemistry to afford **162**. Removal of the protecting group on the N-terminal nitrogen followed by acylation with pyrazine carboxylic acid **163** gives the amide (**164**). Reaction of intermediate **164** with excess isobutyl boronic acid leads to the free acid by transfer of the pinanediol protecting group to the reagent. Thus, **bortezomib** (**165**) is obtained,[24] which is a drug recently approved for treatment of multiple myeloma.

A drug that inhibits HIV by binding with the specific receptor sites on the immune system cells used by the virus to entry into the cells, maraviroc, was discussed in the preceding chapter. The chemical structure of the inhibitor **aplaviroc (179)**, which acts by the same mechanism, is, however, quite different from the drug discussed in Chapter 5. An enantioselective synthesis for the key starting amino acid starts with the oxidation of the double bond in cyclohexyl acrylic ester (**166**) with potassium osmate to afford the cis diol (**167**). Reaction of intermediate **167** with sulfuryl chloride affords the cylic sulfate ester (**168**). Treatment of intermediate **168** with sodium azide leads to attack at the carbon bearing the carboxylate group with inversion of configuration (**169**). Catalytic reduction followed by reaction of the thus-formed amine with *tert*-butyloxycarbonic anhydride leads to intermediate **170** as a single diastereomers.[25]

In a three component synthesis, the amide **171**, obtained from ester **170** and benzyl isocyanide, are reacted with the piperidone (**172**). The product from this transform consists of the addition product (**173**), where amide nitrogen in **170**, as well as that on the isocyanide has added to the carbonyl group on the piperidine. Treatment of adduct **173** with strong acid hydrolyzes the urethane function on the *t*-BOC protecting group leaving behind the primary amine (**174**). Intermediate **174** is not isolated, but undergoes displacement of the benzyl amine function in an intramolecular amide exchange to form the diketo pyridazine (**175**). Hydrogenation of **176** removes the benzyl protecting group to reveal the free piperidine (**177**).

Reductive amination of **177** via the aldehyde group in the phenoxy acid **178** affords **179**.[26]

At least several weeks administration of virtually all antidepressant drugs, be they the classical tricyclic compounds or the newer selective serotonin reuptake inhibitors (SSRI), is required before patients see relief from their symptoms. A recently developed SSRI whose structure departs markedly from existing agents appears to differ from other agents, in that it appears to act in a much shorter time. The final step in the synthesis of this agent **elzasonan (182)** comprises aldol condensation of the benzaldehyde (**180**) with the thiamorpholine (**181**).[27]

REFERENCES

1. F.F. Wagner, D.L. Comins, *J. Org. Chem.* **71**, 8673 (2006).

2. I.W. Davies et al., *J. Org. Chem.* **65**, 8415, (2000).

3. K.H. Andersson, M. Norlander, R.C. Westerlund, U.S. Patent 5,856,346 (1999).

4. A. Mattson, C. Svensson, K. Thornblom, C. Odman, U.S. Patent 6,350,877 (2002).

5. For a discussion of the stereochemistry in a related system see C.J. Barnett, C.R. Copley-Merriman, J. Maki, *J. Org. Chem.* **54**, 4795 (1989).

6. L.A. Sobrera, J. Castaner, *Drugs Future* **28**, 826 (2001).

7. D.M. Armistead, J.O. Saunders, J.S. Boger, U.S. Patent 5,620,971 (1997).

8. K.S. Fors, R.F. Heier, R.C. Kelly, W.R. Perrault, N. Wicnienski, *J. Org. Chem.* **63**, 7348 (1998).

9. Y.W. Hong, Y.N. Lee, H.B. Kim, U.S. Patent 6,252,076 (2001).

10. D.W. Ludovici et al., *Bioorg. Med. Chem. Lett.* **11**, 2235 (2001).

11. M. Ubasawa, S. Yuasa, U.S. Patent 5,604,209 (1997).

12. M. Watanabe, H. Koike, T. Ishiba, T. Okada, S. Seo, K. Hirai, *Bioorg. Med. Chem. Lett.* **5**, 437 (1995).

13. R. Masciardi, U.S. Patent, 5,773,449 (1998).

14. H. Riechers et al., *J. Med. Chem.* **39**, 2123 (1996).

15. K. Yamada, K. Matsuki, K. Omori, K. Kikkawa, U.S. Patent 6,797,709 (2004).

16. M. Hirohashi, M. Kido, Y. Yamamoto, Y. Kojima, K. Jitsikawa, S. Fujii, *Chem. Pharm. Bull.* **41**, 1498 (1993).

17. D. Russowsky, R.F.S. Canto, S.A.A. Sanches, M.G.M. D'Oca, A. de Fatima, R.A. Pilli, L.K. Kohn, M.A. Antonio, J.E. de Carvalho, *Bioorg. Chem.* **34**, 173 (2006).

18. C.H. Tann, P.R. Brodfuehrer, S.P. Brundidge, C. Sapino, H.G. Howell, *J. Org. Chem.* **50**, 3644 (1985).

19. Anon. *New York Times*, March 20, 2004.

20. J.R. Mccarthy, D.P. Mathews, J.S. Sabol, J.R. McConnell, R.E. Donaldson, R. Doguid, U.S. Patent 5,760,210 (1998).

21. B.R. Belleau, C.A. Evans, H.L.A. Tee, *Tetrahedron Lett.* **33**, 6949 (1992).

22. L.S. Jeong et al., *J. Med. Chem.* **36**, 181 (1993).

23. G. Van Lommen, M. De Bruyn, M. Schroven, W. Verschueren, W. Janssens, J. Verrelst, J. Leysen, *Bioorg. Med. Chem. Lett.* **5**, 2649 (1995).

24. J. Adams et al., *Bioorg. Med. Chem. Lett.* **8**, 333 (1998).

25. M. Alonso, F. Santacana, L. Rafecas, A. Riera, *Org. Proc. Resea. Dev.* **9**, 690 (2005).

26. J.A. McIntire, J. Castaner, *Drugs Future* **29**, 677 (2002).

27. J.P. Rainville, T.G. Sinay, S.W. Wallinsky, U.S. Patent 6,608,195 (2003).

CHAPTER 7

FIVE-MEMBERED HETEROCYCLES FUSED TO ONE BENZENE RING

1. COMPOUNDS WITH ONE HETEROATOM

A. Benzofurans

The presence of two iodine atoms on one of the aromatic rings on venerable antiarhythmic agent amiodarone may indicate that it was originally conceived as an agent for treating various thyroid anomalies. A recent halogen-free benzofuran that shares many structural features with its predecessor shows equal or even superior activity in controlling arrythmias. The synthesis starts with an unusual scheme for building the furan ring. Reaction of the benzyl bromide (**1**) with triphenylphosphine leads to the phosphonium salt (**2**). Treatment of the salt with valeryl chloride in the presence of pyridine results in acylation on the now highly activated benzylic carbon. That product (**3**) cyclizes to the benzofuran (**4**) on heating with expulsion of triphenylphosphine. Friedel–Crafts acylation of **4** with anisoyl chloride in the presence of stannic chloride proceeds on the electron-rich furan ring to afford **5**. The nitro group is next reduced with stannous chloride to give amine **6**. That newly formed amine is converted to the corresponding sulfonamide (**7**) by means of methanesulfonyl chloride. Reaction of **7** with boron tribromide cleaves the methyl ether leading to the free phenol (**8**). Alkylation of **8** with chloroethyldibutylamine in the

The Organic Chemistry of Drug Synthesis, Volume 7. By Daniel Lednicer
Copyright © 2008 John Wiley & Sons, Inc.

presence of mild base would then lead to the antiarrythmic agent dronedarone (**9**).[1]

A benzoisofuran illustrates the breadth of the structures that demonstrate SSRI antidepressant activity. Condensation of the phthalide (**10**) with the Gignard reagent (**11**) from 4-bromofluorobenzene leads to addition to the carbonyl group to afford the ring-opened hydroxyketone (**12**). Addition of a single organometallic group to the phtalide may be attributable to the stability of the first-formed addition complex. The benzylic hydroxyl group is next converted to its *t*-BOC derivative **13**. Condensation of that intermediate with the Grignard reagent from 3-chloropropyl dimethylamine **14** leads to the addition product (**15**). The reason for specificity for the ketone over the carbonate may be due to the sterically crowded condition about the latter. Treatment with acid then removes that protecting group. Reaction of the free alcohol with methanesulfonyl chloride forms the mesylate of the primary alcohol. Exposure to triethylamine leads to displacement of that function by the adjacent tertiary hydroxyl group and thus closure to the isobenzofuran ring (**16**). The phenolic hydroxyl is then converted into a leaving group, in this case a triflate, by acylation with trifluoromethyl sulfonyl chloride. Reaction of this last with sodium cyanide, in the presence of triphenylphosphine : palladium and cuprous iodide, replaces the triflate by cyanide. Thus, the antidepressant compound **citalopram** is obtained (**17**).[2]

B. Indoles

A subclass of serotonin receptors dubbed 5-HT$_4$ are involved in regulation of intestinal motility. The partial agonist **tegaserod** (**22**) has proven useful in treating conditions characterized by decreased motility, such as irritable bowel syndrome. The drug is interestingly prescribed specifically for use in women, since results from clinical trials indicated poor response in men. Alkylation of thiosemicarbazone (**18**) with methyl iodide proceeds on sulfur to the thioether (**19**). Treatment of intermediate **19** with pentyl-1-amine leads to addition of the amine to the thioether function followed by elimination of methyl mercaptan to afford the guanidine (**20**). Reaction of **20** with the indole aldehyde (**21**) affords (**22**).[3]

Much of the damage that results from stroke is attributable to increases of glutamate or *N*-methyl-D-aspartate (NMDA), which follow the local anoxia that accompany this event. Agents that inhibit the activity of the resulting abnormal levels of these neurotransmitters would, at least in theory, protect nerves. The NMDA antagonist **gavestinel** (**28**), has a powerful neuroprotective in animal models of stroke. This finding seems not to have held up in the clinic.[4] Condensation of 2,5-dichlorophenylhydazine (**23**) with ethyl pyruvate affords the hydrazone (**24**). Treatment of this intermediate with phosphoric acids leads to rearrangement to the indole (**25**). The indole is treated with *N*-methyl-*N*-phenylformamide in a classic Vilsmeyer reaction. This reaction introduces a formyl group on the remaning open position on the fused pyrrole ring (**26**). Condensation of **26** with the ylide from the amide phosphonium salt leads to the chain extended product **27**. Saponification of the ester group then affords **28**.[5]

Oglufanide (**31**), at one time called thymogen, is a dipeptide isolated from calf thymus. The imunnomodulatory properties of both the natural product and the subsequent synthetic version have been extensively studied as agents that enhances immune function. The compound currently is undergoing clinical trials in patients infected with the hepatitis C virus. In the absence of specific references one may imagine that this dipeptide is obtained using standard peptide chemistry where tryptophan **29** with a protected carboxylic acid is condensed with suitably protected glutamate, (**30**). Removal of the protecting groups would the afford (**31**).

The substrate arachidonic acid, which often leads to formation of inflammatory prostaglandins, is stored in tissues as one of a number of phospholipids; these compounds, as the name indicates, comprise complex phosphate-containing esters. The antiinflammatory corticosteroids inhibit the action of the enzyme, phospholipase A_2, that frees arachidonic acid. The many undesired effects of those steroids has led to the search for non-steroidal inhibitors of that enzyme. A highly substituted indole derivative has shown good activity as a phospholipase A_2 inhibitor. Alkylation of the anion from treatment of indole (**32**) with benzyl chloride affords the corresponding *N*-benzylated derivative (**33**). The methyl ether at the 4 position is then cleaved by means of boron tribromide to yield **34**. Alkylation of the enolate from reaction of the phenol with sodium hydride with *tert*-butylbromoacetate affords the corresponding

O-alkyl product (**35**). Reaction of intermediate **35** with oxalyl chloride proceeds on the only open position of the heterocyclic ring to yield the acylated derivative (**36**) as its acid chloride. Treatment of **36** with ammonia gives the corresponding amide. The *tert*-butyl ester is then cleaved with acid to afford the phospholipase inhibitor **varespladib** (**37**).[6]

Retinopathy, that is, diminishing vision to the point of eventual blindness, is one of the many complications that results from uncontrolled levels of blood sugar that characterize diabetes. The elevated blood sugar that characterize the disease stimulate the enzyme aldose reductase that catalyzes an alternate pathway for glucose metabolism. The sorbitol end-product from this route tends to accumulate in the ocular lens as opacities, a process that leads to increasingly reduced vision and finally total blindness. There is also some evidence that an effective inhibitor would find use in fending off diabetic neuropathy. Though many aldose reductase inhibitors have been identified over the years, few if any have yet achieved regulatory approval. In a convergent synthesis for a recent entry, the amino group in the aniline (**38**) is first acylated with acetic anhydride. Treatment

of the amide (**39**) with tetraphosphorus decasulfide replaces the amide oxygen with sulfur (**40**). Reaction of **40** with sodium hydride produces the sulfide anion from the enol form of the thioamde. This compound displaces the adjacent ring fluorine atom to form a benzothiazole (**41**). Base hydrolysis opens the ring to afford the intermediate mercapto aniline, the sequence in effect having delivered sulfur to the ortho position. In the convergent arm treatment of the anion from sodium hydride and indole (**42**) with ethyl bromoacetate gives the alkylated product (**43**). Condensation of the mercaptoaniline (**44**) with the indole (**43**) in the presence of strong acid couples the two moieties via a new thiazole ring (**45**). The reaction can be rationalized by assuming addition of sulfur to the nitrile as the initial step. Addition–elimination of the aniline nitrogen to the resulting imine then closes the ring. Saponification of the ester on the pendant acetate then affords **lidorestat** (**46**).[7]

Receptors for the estrogens, estrone and estradiol, are characteristically indiscriminate in terms of the structures of the compounds that they will bind. The estrogen antagonists show similarly loose structural requirements for binding (see, e.g., fluvestant in Chapter 2, ospemifene in Chapter 3, lasoxifene in Chapter 4 acolbifene in Chapter 8, and arzoxifene (**141**). An indole provides the nucleus for the estrogen antagonist **bazedoxifene** (**55**); not only the ring system, but also the connectivity of the benzene ring that carries the basic ether differs from earlier compounds. The convergent scheme starts with an unusual method for building an indole. Thus, reaction of the aniline (**47**) with the bromo acetophenone (**48**) in the presence of triethylamine leads to the benzo heterocycle (**49**) in a single step. The aromatic alkylation step required to close the five-membered ring is likely made possible by the high electron density in that ring due the presence of both the ether and the amine functions. Construction of the second ring begins with the alkylation of the phenolic hydroxyl in **50** with ethyl bromoacetate to yield the ether (**51**). The benzylic hydroxyl is then replaced by chlorine by means of thionyl chloride. Condensation of the anion from the indole with the benzylic chloride (**52**) introduces the remaining benzene ring (**53**). The ester on the pendant side chain is next reduced to the corresponding alcohol using lithium aluminum hydride. The terminal hydroxyl is then replaced by bromine in a reaction with carbon tetrabromide in the presence of triphenylphosphine to afford intermediate **54**. Alkylation of **54** with azepine completes construction of the side chain. Catalytic hydrogenation removes the benzyl protecting groups, uncovering the phenolic hydroxyl groups. This affords the estrogen antagonist bazedoxifene (**55**).[8]

The "triptan" class of 5-HT$_3$ serotonin antagonists have found extensive use as drugs for treating migraine headaches. The synthesis of one recent example starts with reaction of the benzylsulfonyl chloride (**56**) with pyrrolidine to afford the sulfonamide (**57**). Hydrogenation of this intermediate reduces the nitro group to the corresponding aniline (**58**). This function is then transformed to a hydrazine by the conventional scheme involving conversion to the diazonium salt by reaction with nitrous acid followed by reduction of this last compound with stannous chloride. Condensation of this product (**59**) with 4-chlorobutyraldehyde gives the hydrazone. Fischer indolization of this last derivative by means of acid affords the indole (**60**). Chlorine on the side chain is then replaced by a primary amine (**61**). Reaction of **61** with formaldehyde and sodium borohydride converts the amine to its *N,N*-dimethyl derivative. Thus, the 5-HT$_3$ antagonist **almotriptan** (**62**) is obtained.[9]

The same general approach is used for the somewhat more complex "triptan" **avitriptan** (**71**). Synthesis of the substituted pyrimidine for the side chain starts with the condensation of dimethylglyoxylate (**63**) with guanidine to afford the pyrimidol (**64**). The hydroxyl group is then replaced by a leaving group, in this case chlorine, by reaction with phosphorus oxychloride. The product from this reaction (**65**) is then alkylated with *N*-carbethoxy piperazine to afford **66**. Treatment of **66** with acid removes the protecting group to afford piperazine **67**, an intermediate suitable for alkylation. Preparation of the indole portion of the molecule begins with the two-step conversion (diazotization then reduction) of the aniline (**68**) to the corresponding hydrazine (**69**). Heating this intermediate with ω-chlorovaleraldehyde, itself prepared in two steps from tetrahydropyran in the presence of acid, leads directly to the indole (**70**), which has in place the requisite side-chain link. Subsequent alkylation of the piperazine (**67**) with **70** leads to the serotonin antagonist **71**.[10]

The highly substituted indole, **ecopladib** (**82**), which shares many structural elements with the phospholipase inhibitor varespladib (**37**), shows similar biological activity. Alkylation of hydroxy benzoate (**72**) with the dimethyl acetal from bromoacetaldehyde (**73**) affords the ether (**74**). Acid-catalyzed reaction of this intermediate with 2-methyl-5-chloroindole (**75**) in the presence of triethylsilane leads in effect to condensation of the acetal with the activated 3 position on the indole ring to afford **76**. The nature of the reduction of the aldehyde carbon is not immediately apparent. Alkylation of the anion on nitrogen from reaction of the indole with sodium hydride and bromodiphenylmethane then adds the third substituent to the fused heterocyclic ring (**77**). The methyl group at the 2 position on this ring is next oxidized with NBS and benzoyl peroxide to yield aldehyde **78**. Reaction of **78** with nitromethane in the presence of ammonium acetate extends the chain by one carbon (**79**). Treatment with zinc amalgam and acid reduces both the nitro group and the double bond to afford the derivative with an ethylamine side chain (**80**). Acylation of that basic nitrogen with the benzylsulfonyl chloride **81** followed by saponification of the methyl ester affords the phospholipase inhibitor **82**.[11]

C. Indolones

As noted earlier tyrosine kinases play a major role in cell proliferation tyrosine kinases comprises a major theme in the search for antitumor agents. Solid tumors are highly dependent on the growth of new blood vessels, a process termed neoangiogenisis, which provide oxygen and nutrients to new tissue masses. The tyrosine kinase inhibitor **semaxanib** (**86**) has shown promising early activity against solid tumors; this compound inhibits neoangiogenisis and also shows antimetastatic activity. Villsmeyer-type reaction of 3,5-dimethylpyrrole (**83**) affords the corresponding carboxaldehyde (**84**). Condensation of **84** with indolone proper (**85**) in the presence of base affords **86**.[12]

The synthesis of a structurally somewhat more complex indolone tyrosine kinase inhibitor starts with the construction of the pyrrole ring. Reaction of *tert*-butyl acetoacetate (**87**) with nitrous acid leads to nitrosation on the activated methylene carbon. This reaction introduces the nitrogen atom that will appear in the target pyrrole. Condensation of **88** with

ethyl acetoacetate (**89**) completes formation of the pyrrole ring (**90**). The strategy depends on the presence of carboxyl groups at the 2 and 4 positions bearing esters with different reactivity. Thus, treatment of the diester (**90**) with aqueous acid leads to hydrolytic decarboxylation of the carboxyl adjacent to the ring nitrogen (**91**). Reaction of this intermediate with methyl orthoformate in strong acid then introduces a formyl group at the only open position on the ring (**92**). Condensation of this aldehyde with indolone **85**, as above, leads to formation of the analogue of **86**, which in addition carries a carboxyl group on the pyrrole ring (**93**). Saponfication then affords the free acid (**94**). Reaction of this compound with *N,N*-diethyl-ethylenediamine gives the corresponding amide. Thus, the tyrosine kinase inhibitor **sunitinib**, (**95**) is obtained.[13]

The neurotransmitter dopamine is closely involved in many aspects of the central nervous system (CNS). It was demonstrated post-facto that the majority of antipsychotic agents owe their efficacy by blocking dopamine receptors in the brain. Parkinson's disease is conversely traceable to a deficiency of dopamine. Most treatments for that disease involve administration of compounds that make up for that deficiency. The indolone **aplindore** (**106**), acts as a partial agonists at the subclass of dopamine receptors associated with Parkinson's. The drug is currently in the clinic for that indication. The compound also interestingly shows some promise for treating "restless leg syndrome". The synthesis begins with alkylation of the free phenol in catechol (**96**) with allyl bromide. The methyl ether in **97** is then cleaved with strong base, a reaction specific for methyl ethers, to afford the phenol (**98**). The newly formed hydroxyl is then alkylated with chiral glycidyl tosylate **99**, to afford the glycidic ether (**100**). Heating a solution of that compound leads first to Claisen rearrangement where the allyl group moves over to the benzene ring to form transient intermediate **101**. The phenol formed as a result of the rearrangement next opens the oxirane to form the dioxin ring (**102**). Reaction of this last compound with carbon tetrabromide in the presence of triphenylphosphine replaces the exocyclic hydroxyl group by bromine (**103**). Oxidation with permanganate next cleaves the double bond in the pendant allyl group to afford the carboxylic acid (**104**). Catalytic hydrogenation of this intermediate serves to reduce the nitro group to the corresponding amine. This product, which is not isolated, cyclizes to afford an amide an thus an indolone ring **105**. Displacement of bromine on the short side-chain by benzylamine completes the synthesis of **106**.[14]

Many different strategies have been investigated to develop agents that will avoid injury to the nervous system that accompanies stroke. The recently developed agent **flindokalner** (**112**), is designed to open the so-called "maxi" potassium channels that enhance an endogenous neuroprotective mechanism. The first step in an enantioseletcive preparation of this compound comprises chlorination of the homosalicylate (**107**) by means of sulfuryl chloride. Fischer esterification in methanol affords the

methyl ester (**108**). Treatment with potassium hexamethylsisilazide leads to formation of the benzylic anion. Reaction of this anion with the nitro compound leads to nucleophilic aromatic displacement of the ring fluorine atom by the benzylic anion and formation of intermediate **109**. The presence of the two electron-withdrawing groups in the nitro compound facilitate this reaction. The anion is of course not isolated, and is next quenched with *N*-fluorobenzenesulfonimide to afford **110** as a mixture of enantiomers. The ester group in this mixture is then saponified; the isomers of the resulting carboxylic acid are next separated by way of their (*S*)-benzylamine salts. The isomer that corresponds to the (*S*)-isomer is then treated with sodium dithionate in order to reduce the nitro group to the corresponding amine (**111**). Treatment of this ortho amino acid with acid closes the ring to an indolone. Thus, the neuroprotective agent **112** is obtained.[15]

No biologically active compound has arguably had quite the effect on the governmental regulations that affect the pharmaceutical industry as has thalidomide. The profound malformations caused in offspring of women taking this nonbarbiturate sedative caused reappraisal of government oversight of pharmaceuticals in most of western Europe. In the United States, the event provided the final push for a major revision of the Food and Drug Laws that govern the FDA. The low-level research that continued on this drug, in spite of its ill repute, unexpectedly showed that the compound affected immune function. The drug was, for example, recently approved by the FDA for treatment of complications from leprosy; it has also been investigated as an adjunct for treating some malignancies. Recent research on related compounds has revealed a series that inhibits tumor necrosis factor (TNF-α). Synthesis of the aminosuccinimide moiety starts by cyclization of carbobenzyloxy glutamine (**113**) by treatment with carbonyl diimidazole (CDI). Catalytic hydrogenation of the product removes the protecting group. The aromatic moiety

of the target compound is prepared by free radical bromination of the methyl group in the benzoic acid (**115**) to give the bromomethyl derivative (**116**). Condensation of this last with the succinimide (**114**), leads to isoindolone (**117**). A second hydrogenation step reduces the nitro group to an aniline to afford **lenalidomide (118)**.[16]

D. Miscellaneous Compounds with One Heteroatom

Growth hormone has long been used for treating children whose retarded growth was traceable to deficient levels of the polypeptide. This high molecular weight compound that was used for treating deficiencies was originally obtained from cadaver pituitary glands. Growth hormone produced by recombinant technology became available at almost the exactly time that it was recognized that the cadaver derived drug could potentially carry the prions that cause fatal Creutzfelt–Jacob disease. The recombinant derived material has as a result completely replaced the pituitary-derived hormone. The cost of this drug, which can run as high as $170 per day of treatment, has encouraged the search for small molecules. The dihydroindole **ibutamoren (128)**, achieves much the same purpose as the polypeptide; the drug does so by stimulating secretion of growth hormone from the pituitary gland. This agent has shown promising results in the clinic. The reaction sequence for its preparation begins with acylation of the amino group in spiroindoline (**119**) with carbobenzyloxy chloride to afford **120**. The methyl group on the piperidine ring in this product is then removed using a modern version of the von Braun reaction (**121**). Acylation of this product with the *t*-BOC derivative of benzyloxyserine (**122**) using standard peptide-forming condition leads to the amide (**123**). Treatment of that intermediate with trifuoroacetic acid removes the *t*-BOC protecting group to reveal the free primary amine (**124**). The peptide-like side chain is next extended by acylation with *t*-BOC protected aminoisobutyric acid (**125**) to give **126**. Catalytic hydrogenation next removes the Cbz group that has protected the indoline nitrogen through the preceding steps (**127**). Reaction of **127** with methanesulfonyl chloride gives the corresponding sulfonamide. Trifluroacetic acid then removes the remaining *t*-BOC group to afford **128**.[17]

Compounds whose structures include a quinone moiety have been intensively investigated as potential antitumor agents. At least two quinones, mitomycin C and diaziquone, that have found their way to the clinic. These compounds in addition include a reactive aziridine ring. A recent entry that incorporates both those features, **apaziquone** (**135**), also known as EO9, may be viewed as an oxidized indole. In the key reaction of a succinct synthesis to this agent, quinone **129** is allowed to react with

the *N*-methylenamine from 4-methoxyacetoacetate **130** to form the fused pyrrole ring in a single step. The reaction can be rationalized by assuming that the first step involves displacement (or addition–elimination) of bromine by the basic nitrogen on the eneamine; conjugate addition of the active methylene group to the quinone then closes the ring to afford the observed product (**131**). Treatment with DDQ next oxidizes the exocyclic methoxymethylene group to the corresponding aldehyde (**132**). Condensation of this product with the phosphorane from Emmon's reagent results in the homologated product **133**; the added salt assures the trans configuration of the newly introduced double bond. In the next step, reduction with DIBAL converts both esters to the corresponding alcohols (**134**). Reaction of this last intermediate with aziridine results in displacement (or again, addition–elimination) of the ring methoxy group. Thus, the antitumor agent **135** is obtained.[18]

In yet another illustration of the breadth of the SAR of estrogen antagonists, the carbonyl group in the estrogen antagonist raloxifene can be replaced by ether oxygen. Reaction of the benzothiophene (**136**) with bromine leads to the derivative (**136**) halogenated at the 3 position (**137**). The ring sulfur atom is next oxidized with hydrogen peroxide in order to activate that bromine toward displacement (**138**). Reaction of this intermediate with the anion from treatment of the phenol (**139**) with sodium hydride displaces bromine, and in a single step introduces the ring that carries the requite basic ether. The sulfoxide function is next reduced to the sulfide oxidation state by means of lithium aluminum hydride (**140**). Scission of the methyl ethers with boron tribromide completes the synthesis of the estrogen antagonist **arzoxifene** (**141**).[19]

2. FIVE-MEMBERED RINGS WITH TWO HETEROCYCLIC ATOMS

A. Benzimidazoles

Thrombin is an essential intermediate in the formation of blood clots. Inhibitors of this factor would prove useful in treating and preventing inappropriate clot formation that potentially leads to stroke and heart attacks. Reaction of the carboxylic acid (**143**) with thionyl chloride leads to the corresponding acid chloride (**144**). Treatment of that intermediate with the substituted pyridyl amine (**142**) leads to the amide (**145**). Catalytic hydrogenation of the product reduces the nitro group to the primary amine (**146**). Condensation of this ortho diamine with the carboxylic acid (**147**) in the presence of carbonyl diimidazole then forms the imidazole ring (**148**). This reaction proceeds via the amide formed with the primary amine followed by replacement of the amide carbonyl oxygen by the adjacent amine. Reaction of the product with ammonium carbonate leads to addition of ammonia to the nitrile to form an amidine. Saponification of the side-chain ester affords the thrombin inhibitor **dabigartan (149)**.[20]

Thioazolidinediones have by now achieved a major role in the treatment of Type II, also called adult onset diabetes. A recent example includes a benzimidazole moiety as part of the structure. One of the syntheses for this compound starts by nucleophilic aromatic displacement of fluorine in p-fluorobenzaldehyde (**151**) by the anion from treatment of the hydroxymethylbenzimidazole (**150**) with strong base. The product (**152**) is then treated with thiazolidinedione proper (**153**) in the presence of base. The anion from the methylene group in **153** adds to the aldehyde; the transient alcohol dehydrates to form the observed product **154**. Catalytic hydrogenation of that product then affords **rivoglitazone (155)**.[21]

Purines in which the sugar moiety is replaced by an abbreviated open-chain surrogate, for example, acyclovir, comprise an important group of antiviral drugs. Antiviral activity is retained when the pyrimidine ring in the heterocyclic part of those molecules is replaced by benzene, though the example at hand includes a normal sugar. Thus, coupling of the benzimidazole (**156**), with the tetracetyl ribofuranoside (**157**) in the presence of *N,O*-bis(trimethylsylil acetamide) affords the glycosilated benzimidazole (**158**). Treatment on this intermediate with isopropylamine leads to displacement of the bromine on the imidazole ring by isopropyl amine (**159**). Saponification with aqueous sodium carbonate removes the acetyl protecting groups to afford the antiviral agent **maribavir** (**160**).[22]

Antidepressant drugs, whether they belong to the tricyclic or newer SSRI series, do not as a rule become effective until ~2 weeks after treatment has started. The recent 5HT antagonist **fibanserin** (**166**), departs from that pattern in that it starts to elevate patient's mood within a short time after the treatment has started. Reaction of benzimidazolone (**161**) with benzoyl chloride affords the singly protected derivative **162**. Alkylation of the anion prepared from this intermediate with sodium hydride with 1,2-dichloroethane adds the linking chain for attaching the next large moiety (**163**). The benzoyl group on the other imidazolone nitrogen is then hydrolyzed in the presence of acid (**164**). Displacement of the terminal chlorine with the arylpiperazine **165** affords the alkylation product **166** and thus **166**.[23]

B. Miscellaneous Compounds

Many new antipsychotic compounds that have been introduced over the years that showed decreased side effect in initial trials. The promise of such "atypical" agents often did not stand up with chronic use. The most recent atypical drug is a mixed agonist–antagonist at dopamine receptors. The partial agonist activity should in theory avoid the effects of excessive blockade. The compound at hand also acts on 5-HT receptors; this should help patients who suffer from bipolar disorder in the depressed phase of the disease. Alkylation of the nitrogen on the piperazine (**167**) with the 3-bromobenzyl mesylate (**168**), obtained by reaction of the corresponding alcohol with methanesulfonyl chloride, affords the intermediate (**169**). The additional benzene ring is added by Suzuki cross-coupling with phenyl-boronic acid. Thus, the antipsychotic compound, **bifeprunox** (**170**).[24]

The search for compounds that have a neuroprotective effect following ischemic stroke is a recurrent theme in this chapter as well as in Chapter 3. This search for effective drugs has ranged over a variety of biochemical

mechanisms, as well chemical structures. A benzothiazole moiety forms part of a compound that acts as a neuroprotectant in animal models. The (S) enantiomer of this compound is far more active than its antipode both *in vitro* and *in vivo*. One of several methods for preparing the substituted thiazole[25] comprises displacement of the mesylate group in benzothiazole (**173**) by the free amine in protected aminopiperidine (**172**). Hydrolysis then removes the protecting group to reveal the free piperidine amine (**175**). In a convergent step, the anion from difluorophenol (**176**) is reacted with chiral epichlorohydrin (**177**) to afford **178**. This reaction proceeds with overall retention of configuration whether initial attack is on chlorine or the epoxide. Reaction of this intermediate with the piperidine (**175**) affords the product with a linkage reminiscent of a β-blocker. This completes the synthesis of the (*S*)-isomer of **lubeluzole** (**179**).[26]

REFERENCES

1. C. Mellin, U.S. Patent 5,854,282 (1998). Note this patent describes the chemistry up to phenol **8**. The alkylation step is conjectural though well precedented.
2. Anon, *Drugs Future* **25**, 620 (2000).
3. K.-H. Buchheit, R. Gamse, R. Giger, D. Hoyer, F. Klein, E. Klopner, H.-J. Pfannkuche, H. Mattes, *J. Med. Chem.* **38**, 2331 (1995).
4. E.C. Haley et al., *Stroke* **36**, 1006 (2005).
5. R.D. Fabio, A. Cugola, D. Donati, A. Ferari, G. Gariraghi, E. Ratti, D.G. Trists, A. Reggiani, *Drugs Future* **23**, 61, (1998).
6. S.D. Draheim et al., *J. Med. Chem.* **39**, 5159 (1996).
7. M.C. Van Zandt et al., *J. Med. Chem.* **48**, 3141 (2005).
8. C.P. Miller, H.A. Harris, B.S. Komm, *Drugs Future* **27**, 117 (2002).
9. Anon, *Drugs Future* **24**, 367 (1999).
10. P.R. Brodfuehrer et al., *J. Org. Chem.* **62**, 9192 (1997).
11. J.C. McKew, S.Y. Tam, K.L. Lee, L. Chen, P. Thakker, F.-W. Sum, M. Behnke, B. Hu, J.D. Clark, U.S. Patent 6,797,708 (2004).
12. L. Sun, N. Tran, H. App, P. Hirth, G. McMahon, C. Tan, *J. Med. Chem.* **41**, 2588 (1998).
13. L. Sun et al., *J. Med. Chem.* **46**, 1116 (2003).
14. G.P. Stack, R.F. Mewshaw, B.A. Bravo, Y.H. Kang, U.S. Patent 5,962,465 (1999).
15. P. Hewawasam et al., *Bioorg. Med. Chem. Lett.* **12**, 1023 (2002).
16. G.W. Muller, R. Chen, S.-Y. Huang, L.G. Corral, L.M. Wong, R.T. Patterson, Y. Chen, G. Kaplan, D.I. Stirling, *Bioorg. Med. Chem. Lett.* **9**, 1625 (1999).
17. D.C. Dean et al., *J. Med. Chem.* **39**, 1767 (1996).

18. E. Comer, W.S. Murphy, ARKIVOC, 286 (2003).

19. A.D. Palkowitz et al., *J. Med. Chem.* **40**, 1407 (1997).

20. N.H. Hauel, H. Nar, H. Priepe, U. Ries, J.-M. Stassen, W. Wienen, *J. Med. Chem.* **45**, 1757 (2002).

21. For general method see, T. Fujita, K. Wada, M. Oguchi, H. Yanagisawa, K. Fujimoto, T. Fujiwara, H. Horikoshi, T. Yoshioka, U.S. Patent 5,886,014 (1999) and T. Fujita, T. Yoshioka, T. Fujiwara, M. Ogichi, H. Yanagisawa H. Horikoshi, K. Wada, K. Fujimoto, U.S. Patent 6,117,893 (2000).

22. S.D. Chamberlain, G.W. Koszalka, J.H. Tidwell, N.A. Vandraanen, U.S. Patent 6,204,249 (2001).

23. G. Bietti, F. Borsini, M. Turconi, E. Giraldo, M. Bignotti, U.S. Patent 5,576,318 (1996).

24. R.W. Feenstra, J. de Moes, J.J. Hofma, H. Kling, W. Kuipers, S.K. Long, M.T.M. Tulp, J.A.M. van der Heyden, C.G. Kruse, *Bioorg. Med. Chem. Lett.* **11**, 2345 (2001).

25. R.A. Stokbroekx, M.G.M. Luyckx, F.E. Janssens, U.S. Patent 4,861,785 (1989).

26. R.A. Stokbroekx, G.A.J. Grauwels, U.S. Patent 5,434,168.

CHAPTER 8

SIX-MEMBERED HETEROCYCLES FUSED TO ONE BENZENE RING

1. COMPOUNDS WITH ONE HETEROATOM

A. Benzopyrans

A prominent feature in the majority of estrogen antagonists consists of a pair of aromatic rings disposed on adjacent positions on either an acyclic ethylene or a fused bicyclic ring system; in the latter case, one of those rings is positioned next to the ring fusion. Those two moieties are present in the antagonist **acolbifene (9)**; the structure of this agent, however, departs from previous examples by the fact they occupy the 2,3- rather than 1,2-position of the bicyclic system. This finding may account for the report that this agent is a pure antagonist that, unlike its predecessors, shows no partial agonist activity at estrogen receptors. The synthesis of this agent begins with Fiedel–Crafts acylation of resorcinol (**1**) with 4-hydroxyphenylacetic acid (**2**) to afford the desoxybenzoin (**3**). Reaction with dihydropyran (DHP) leads to formation of the bis(tetrahydro pyrranyl) ether (**4**); the phenolic group adjacent to the ketone does not react as a result of its chelation with that carbonyl group. Condensation of that intermediate with *p*-hydroxybenzaldehyde in the presence of piperidine initially leads to the chalcone (**5**). The phenol then adds to the double bond conjugated with the carbonyl group to afford the benzopyranone

The Organic Chemistry of Drug Synthesis, Volume 7. By Daniel Lednicer
Copyright © 2008 John Wiley & Sons, Inc.

(6). The free phenol group on the newly introduced ring is then alkylated with *N*,4-chloroethylpiperidine to give the basic ether (7). Reaction of 7 with methylmagnesium bromide followed by treatment with acid leads the first-formed alcohol to dehydrate; at the same time the tetrahydropyrranyl groups hydrolyze to reveal the free phenols 8. Separation on a chiral chromatographic column then affords the (S) isomer, 9.[1]

Migraine was a condition that was refractory to treatment until the discovery of the serotonin receptor blocker, such as sumatriptan. This agent was soon followed by several other drugs that acted by the same mechanism. **Tidembersat (13)** a compound more closely related, both in structure and mechanism of action, to antihypertensive benzopyrans that act on potassium channels showed early promise as an agent for treating migraine.[2] Lack of recent references suggest that this activity did not hold up in the clinic. The synthesis that follows[3] is somewhat conjectural as it depends on a sketchy description in a patent. Thus stereoselective epoxidation of the benzopyran (**10**), using, for example, Sharpless conditions, affords the chiral oxirane (**11**). Ring opening that intermediate with ammonia or an equivalent would then afford the trans aminoalcohol (**12**). Acylation with 3,5-difluoro-benzoyl chloride would give **13** as a single diastreomer.

Every cell in a living organism incorporates a time clock that marks its longevity. The cell then dies at the end of its prescribed lifetime. This process, called apoptosis, is disrupted in cancer cells and results in their immortality. One of the approaches to treating neoplasms involves restoring this process. The benzopyran **alvocidib** (**20**), perhaps better known by its previous name **flavoperidol**, has shown promising activity as an agent that restores apoptosis. The scheme below is based on that for the compound in which methyl replaces the pendant chlorobenzene ring.[4] The sequence starts with addition of diborane to the olefin in the tetrahydropyridine (**14**), itself available by addition of an organometallic reagent to *N*-methyl-4-piperidone followed by dehydration of the tertiary alcohol. Oxidation of the hydroborane adduct with a peroxide then affords the hydration product as its cis isomer (**15**). The stereochemistry at that center is then reversed by oxidation to the corresponding ketone followed by reduction with sodium borohydride (**16**). This intermediate is acylated with acetic anhydride in the presence of boron trifluoride. Use of an excess of the latter leads to selective demethylation of the ether adjacent to the newly introduce acyl function (**17**). Claisen condensation of this product with methyl 2-chlorobenzoate gives the β-diketone (**18**). Treatment with acid causes the phenolic oxygen to add to the enone, thus forming the pyran ring by an addition–elimination sequence (**19**). Demethylation of the remaining phenolic ethers, for example, with boron tribromide, would then afford **20**.[5]

The continuing search for better tolerated antipsychotic drugs has led to the examination of subtype dopamine receptors in the hope that selectivity might avoid the side effects that come with indiscriminate receptor blockade. A benzopyran compound that shows preferential binding to the D_4 dopamine receptors showed promising results in the laboratory; this agent, however, subsequently failed in the clinic. The benzopyran nucleus is formed in a single step. Thus, reaction of phenethyl alcohol (**21**) with the acetal (**22**) in the presence of strong acid initially leads to the transient intermediate (**23**) in which one of the methoxy groups has been replaced by the hydroxyl group from the phenethyl reagent. The carbocation formed under the strongly acidic conditions by loss of the remaining methoxyl then attacks the aromatic ring to form a pyran (**24**). Alkylation of the secondary amine in the piperazine (**25**) with the halogen in **24** then affords **26**. This last is resolved to afford **sonepiprazole** (**27**).[6]

B. Quinolines and Their Derivatives

Multidrug resistance poses an increasingly important problem for chemotherapy. Several compounds, zosuquidar (Chapter 4) and biriquidar (Chapter 6), that attempt to overcome this by interfering with the mechanism by which cells extrude drugs were noted earlier. A compound that includes in its structure both a quinoline and a perhydroisoquinoline moiety shares the mechanism of action with its predecessors. The perhydroisoquinoline moiety (**31**) can conjecturally be prepared by alkylation of **28** with the chloroethyl nitrobenzene (**29**). Catalytic hydrogenation of the product (**30**) would then reduce the nitro group to afford intermediate **31**. Acylation of the newly formed amine with the nitrobenzoic acid (**32**) in the presence of cyclohexyl-morpholino-carbodiimide then yields amide **33**. Hydrogenation again converts the nitro group to the corresponding aniline (**34**). Acylation of intermediate **34** with quinoline-3-carboxylic acid (**35**) yield the extended antharanilamide. Thus **tariquidar (36)** is obtained.[7]

As noted in Chapters 4, 5, and 7, compounds aimed at disrupting tyrosine kinases have been intensively studied as potential antitumor compounds. The quinoline-based inhibitor **pelitinib (45)** incorporates a Michael acceptor function in the side chain that can form a covalent bond with a nucleophile on the target enzyme. Such an interaction would result in irreversible inhibition of the target kinase. Reaction of the aniline (**37**) with DMF acetal leads to the addition of a carbon atom to aniline nitrogen in the form of an amidine (**38**). This intermediate is next reacted with nitric in acetic acid to form the nitrated product

(**39**). Condensation of the product from that reaction with ethyl cyano-acetate in the presence of acid affords the enamine (**40**) from displacement of dimethylamine. Heating this product in Dowtherm closes the ring via the ester group to form the cyano quinoline (**41**). The next step invokes one of the standard schemes in heterocyclic chemistry used to transform a hydroxyl to a better leaving group. Treatment of the enol (**41**) with phosphorus oxychloride thus affords the chlorinated derivative (**42**). Reaction of **42** with the halogenated aniline (**43**) leads to displacement of chlorine by the basic nitrogen and formation of **43**. Treatment of that intermediate with iron powder in acetic acid serves to reduce the nitro group that was put in place early in the sequence to the amine (**44**). Acylation of that newly introduced amine with 4-*N,N*-dimethylaminobut-2-enol chloride gives the irreversible kinase inhibitor pelitinib (**45**).[8]

Neurokinins comprise a group of peptides involved in nerve transmission. Specific members of this class of mediators control such diverse functions as visceral regulation, and CNS function. The nonpeptide neurokinin antagonist **talnetant** (**51**), for example, is currently being evaluated for its effect on irritable bowel syndrome, urinary incontinence, as well as depression and schizophrenia.[9] The quinoline portion of this compound is prepared by base-catalyzed Pfitzinger condensation of isatin (**46**) with the methoxy acetophenone (**47**). The methoxy ether in the product (**48**) is next cleaved by means of hydrogen bromide. Amide formation of (**49**) with the chiral α-phenylpropylamine (**50**) affords the antagonist (**51**).[10]

Cholesterol is not absorbed from the intestine as such, it first needs to be esterified. Another enzyme, cholesteryl ester transfer protein (CETP), then completes absorption of cholesterol. Drugs that interfere with the action of these peptides would aid in lowering cholesterol levels by complementing the action of the statins that inhibit endogenous production of cholesterol. The CEPT inhibitor **torcetrapib** (**60**), proved very effective in lowering cholesterol levels in humans; the drug not only lowered low-density lipoproteins (LDL and VLDL) but also raised levels of high density, "good" lipoproteins (HDL). The drug was, however, withdrawn from the market not long after its introduction as a result of serious concerns as to its safety. The synthesis of this compound involves an unusual method for preparing the tetrahydroquinoline moiety. The sequence starts with the reaction of the trifluormethylaniline (**52**) with propanal in the presence of benzotriazole (**53**) to produce the aminal **54**. Condensation of **54** with the vinyl carbamate (**55**) adds three carbon atoms in what may be viewed formally as a 3 + 3 cycloaddition sequence. This yields the tetrahydroquinoline ring (**56**) with expulsion of the benzotriazole fragment. The ring nitrogen is then protected as its ethyl carbamate by acylation with ethyl chloroformate (**57**). The benzyl carbamate function on nitrogen at the 4 position is next removed by reduction with ammonium formate over palladium to afford the primary amine **58**; this compound is then resolved as its dibenzyl tartrate salt to afford the (2R, 4S) isomer. Reductive amination with the bis(trifuoromethyl benzaldehyde) (**59**) in the presence of sodium triacetoxy borohydride. Acylation with methyl chloroformate completes the synthesis of **60**.[11]

As noted previously, anticholinergic agents have undergone something of a renaissance due to their use as drugs to treat urinary incontinence. The synthesis of one of these muscarinic antagonsists starts with the classic

Pictet–Spengler method for building a dihydroisoquinoline ring. Thus, treatment of the benzamide (**61**) from 2-phenethylamine with phosphorus oxychloride probably results in initial formation of a transient enol chloride; this then cyclizes to **62** under reaction conditions. The imine is then reduced with sodium borohydride **63**. Resolution by means of the tartrate salt affords **64** in optically pure form. Acylation of this intermediate with ethyl chloroformate leads to carbamate **64**. Reaction of **64** with the anion from chiral quiniclidol (**65**) interestingly results in the equivalent of an ester interchange. Thus, the anticholinergic agent **solifenacin (66)** is obtained.[12]

Addition of the terpene, farnesol, to cysteine residues near the end of protein chains is a crucial process for transporting some proteins to the intended membrane compartment. This process thus plays an important role in cell proliferation. Inhibitors of the enzyme that catalyzes farnesylation, farnesyl transferase, provide yet one more mechanism for interrupting the multiplication of malignant cells. One of several synthesis of this agent

starts with the acylation of *N*-methylaniline (**68**) with the cinnamoyl chloride (**67**). Treatment of the resulting amide (**69**) with polyphosphoric acid leads to attack of the protonated olefin onto the adjacent benzene ring with formation of the tetrahydroquinolone (**70**). This intermediate is then reacted with 4-chlorobenzoyl chloride in the presence of a Friedel–Crafts catalyst to afford the corresponding ketone. The heterocyclic ring is next dehydrogenated by reaction with bromine. The initially formed quaternary bromide apparent loses hydrogen bromide under reaction conditions to give the unsaturated quinolone (**71**). Treatment of **71** with the anion obtained from *N*-methylimidazole and butyllithium leads to addition of that heterocycle to the ketone. The thus formed carbinol (**72**) is then treated with ammonia. The quaternary carbinol then in essence solvolyses so as to replace the hydroxyl group by a primary amine forming **tipifarnib (73)**.[13,14]

Agents that increase the force of contraction of the heart, often called positive inotropic agents, play an important role in treatment of congestive heart failure. Many of the currently available drugs also increase heart rate, an undesired side effect. A compound that is based on a quinolone is said to increase force of contraction without speeding up the heart rate. The final steps of synthesis on this agent closely parallel those used to prepare β-blockers. Thus, reaction of the carbostiryl (**74**) with epichlorohydrin affords the glycidic ether (**75**). Treatment of this intermediate with the benzylamine (**76**), opens the epoxide to afford the aminoalcohol **torborinone (77)**.[15]

C. Quinolone Antibacterial Agents

Research on the quinolone antibacterial agents crested a decade ago, as indicated by the fact that Volume 5 in this series described the synthesis of no fewer than 11 drugs in this structural class. The level of activity then, not surprisingly, declined so that only four quinolones were described in Volume 6, which was published in 1999. Two of the three quinolones discussed below were actually prepared before that year; their absence in the book is due to the circumstance that they had not yet, for some reason, been assigned nonproprietary names. By way of a reminder, the quinolones act by interfering with the bacterial enzyme, DNA gyrase. An important step in cell replication involves reading the genome in order to generate a corresponding RNA strand. The sheer size of the genome requires that it be cut in order to gain access to the relevant section. Gyrases and related topoisomerases temporarily hold the cut ends during transcription; other enzymes reconnect those ends when the process is complete. Topoisomerase inhibitors cause the formation of covalent bonds so that the genome is frozen in the cut position and as a consequence can no longer function. Cell replication is thus brought to a halt.

The synthesis of one of the agents begins with nucleophilic aromatic displacement of bromine by cyanide in the highly fluorinated compound **78**. Acid hydrolysis of the nitrile (**79**), followed by esterification of the newly formed acid, affords ester **80**. Base-catalyzed condensation of the intermediate with diethyl malonate leads to the tricarbonyl derivative **81**.

This loses one of the carboxylates on heating in the presence of toluene-sulfonic acid to afford the β-ketoester (**82**). Reaction of this intermediate with ethylorthoformate then adds a carbon atom to the activated methylene. Heating that compound with cyclopropylamine in effect exchanges the ethoxy group with the amine to afford enamine (**84**). Treatment **84** with sodium fluoride leads to displacement of one of the ring fluoro groups by the basic nitrogen on the side chain. This step concludes the formation of the quinolone (**85**). Reaction of **85** with 2-methylpiperazine leads to a second displacement of fluorine, in this case by the less hindered of the basic nitrogen atoms on the piperazine. Saponification of the ester then affords **gatifloxacin (86)**.[16]

Replacing one of the protons on the cyclopropyl group by fluorine introduces an element of asymmetry into that moiety. A significant portion of the synthesis of **sitaloxacin (98)** is consequently devoted to the preparation of that substituent in chiral form. Reaction of the chiral auxiliary amino-alcohol (**87**) with phosgene closes the ring to afford the oxazolidone (**88**). This compound is then treated with the methyl acetal from acetaldehyde; the ring amide nitrogen displaces one of the methoxy groups to give the corresponding carbinolamine derivative **89**. Heating **89** leads to loss of methanol to yield the vinyl amine (**90**). Addition of fluoromethylene carbene, generated from fluoromethyl iodide and diethyl zinc leads to formation of the cyclopropyl group (**91**). The reaction proceeds to afford

the chiral cis isomer due to the steering influence of the proximate chiral auxiliary. Catalytic hydrogenation leads to scission of the benzyl–nitrogen and benzyl–oxygen bonds; the transient carbonate disintegrates on work up to afford chiral *cis*-amine (**92**). The scheme then proceeds much as that above though details differ. Thus, reaction of **92** with the ethoxymethylene intermediate (**93**) leads the amine to replace the ethoxy group to afford

94. The cyclization to a quinolone (**95**) in this case is effected with sodium hydride. Treatment of this intermediate with the spiro diamine (**96**) leads to displacement of fluorine and formation of alkylation product **97**. Deprotection by acid-catalyzed cleavage of the *t*-BOC group flowed by saponification yields the quinolone antibacterial agent **98**.[17]

The first few reactions in the preparation of the most recent of this small set of quinolones involves adjustment of the substitution pattern on the central benzene ring. Thus carbonation of the lithio derivative from **99** with carbon dioxide gives the corresponding acid; this acid is then

converted to the methyl ester with diazomethane to yield **100**. The methyl ether is then cleaved by means of boron tribromide to yield **101**. The newly revealed phenol is then alkylated with chlorofluromethane in the presence of base to afford the ether (**102**). Reaction of the ester with sodium azide leads to nucleophilic aromatic displacement of the fluorine atom.

(The reagents used for these last two steps almost guarantee that this is not the route used for larger scale synthesis.) Catalytic hydrogenation then reduces the resulting azide to the corresponding aniline. Saponification of the ester leads to amino acid **103**. The amine is then diazotized and the diazonium salt treated with hydrogen bromide to afford bromo derivative **104**. Condensation of this last intermediate with the magnesium salt from ethyl malonate leads in a single step to the requisite β-ketoester (**105**). The extra carbon required to build the fused ring is added in this case by reaction with DMF acetal (**106**). The methyl enol ether is then displaced as above by cyclopropylamine (**107**). Treatment of the product with base closes the ring to afford the quinolone (**108**). Suzuki cross-coupling of the bromine atom in **108** with the boronic acid from dihydroisoindole **109** leads to the coupling product (**110**). The trityl protecting group on isoindole nitrogen is then removed by treatment with acid. Thus the quinolone **garfenoxacin** (**111**) is obtained.[18]

2. COMPOUNDS WITH TWO HETEROATOMS

A. Benzoxazines

The multidrug regimen currently used to treat HIV infection includes three drugs, a proteiase inhibitor, and two reverse transcriptase inhibitors. In order to try to avoid the development of drug resistance, one of the latter will consist of a modified nucleoside, while the second will be one of several nonnucleosides inhibitors, such as those noted earlier in this volume [capravirine (Chapter 5), etravirine (Chapter 6)]. The structure of the benzoxazine inhibitor **efavirenz** (**121**), which differs significantly from earlier agent points up to the wide structural divergence among this class of antiviral agents. Acylation of *p*-chloroaniline (**112**) with pivaloyl chloride affords the corresponding amide (**113**). Treatment with butyllithium followed by ethyltrifluoroacetate introduces the required trifluoroacetyl group (**114**). Acid hydrolysis then removes the pivaloyl group to afford the free amine (**115**). This function is then protected from reagents in the rest of the sequence by alkylation with *p*-methoxybenzyl bromide (**116**). The key reaction in the sequence involves stereospecific addition of the cyclopropylacetylene moiety. Thus, addition of the lithium acetylide from cyclopropylacetylene to the trifluoromethyl group in **117** in the presence of the substituted ephedrine derivative **118** proceeds with high enantiomeric excess (ee). Reaction of the thus-obtained aminoalcohol (**119**) with phosgene closes the benzoxazine ring (**120**). The methoxybenzyl group is then removed under reductive conditions. This last reaction affords **efavirenz** (**121**) as a single enantiomer.[19]

The structural promiscuity of the estrogen receptors has led to the development of a host of nonsteroidal agonists and antagonist. Nonsteroidal compounds that bind to either progestin or androgen receptors have, on the other hand, proven far more elusive, arguably because of the more rigid structural demands at those sites. A benzoxazine has very recently been found to act as a potent agonist at progesterone receptors both *in vitro* and *in vivo*. Molecular modeling suggests that the benzoxazine moiety in this compound fulfills the role of the steroid AB rings while the pyrrole fulfills the role of the D ring in progesterone.[20] Construction of the benzoxazine starts with the Grignard reaction of anthranilate **122** with methylmagnesium bromide. Treatment of the product (**123**) with carbonyl diimidazole closes the oxazole ring **124**. In a convergent scheme, reaction of N-protected imidazole **125** with butyllithium followed by trimethyl borate affords the boric acid derivative **126**. Condensation of this acid with benzoxazole **124** in the presence of the palladium/triphenyl-phosphine catalyst affords the coupling product **127**. Treatment of **127** with isocyanosulfonyl chloride adds the required cyano function to the pyrrole **128**. The protecting group on the imidazole is then removed by means of sodium ethoxide. The free amine is next methylated by means of methyl iodide in the presence of potassium carbonate. The final step involves conversion of the carbonyl group to its sulfur equivalent. Treatment of **129b** with Lawesson reagent from phosphorus sulfide (P_4S_{10}) and anisole, then affords the nonsteroidal progestin **tanaproget 130**.[20]

B. Quinazolines

Benign prostatic hypertrophy (BPH) comprises an annoying though hardly life-threatening condition that faces many men as they age. As the anti-hypertensive agent prazocin came into widespread use, reports began to accumulate of relief of BPH symptoms by patients who were taking this α-2 sympathetic blocker These reports were later substantiated by formal clinical trials. Detailed pharmacology then revealed that α-2 sympathetic receptors occur in the prostate. The relief from BPH symptoms is now attributed to the blockade of those receptors. This discovery was followed by the introduction of compounds targeted specifically at this new indication. The α-2 blocker **alfuzocin** (**139**), shares the quinazoline moiety with prazocin and some of its later analogues. One arm the convergent synthesis to this agent starts with acylation of amine **131** with the ethoxy-carbonate derivative (**132**) of tetrahydrofuroic acid to afford amide **133**. Hydrogenation then reduces the cyano group to the corresponding primary amine **134**. Preparation of the quinazoline moiety first involves reaction of the dicarbonyl compound **135** with phosphorus oxychloride to afford the dichloro derivative (**136**). Treatment with ammonia results in the displacement of the more labile halogen to give the monoamine (**137**). Reaction of this last intermediate with the future side-chain amine **134** under more forcing conditions, leads to displacement of the remaining ring halogen. Thus the α-2 blocker **139** is obtained.[21]

The particularly good activity against protein kinases of α-aminoquin-azoline derivatives is borne out by their activity against both *in vitro* and *in vivo* models of human tumor. The seven compounds that follow reflect the large amount of research that has recently been devoted to this structural class.

In the first example, nitration of the benzoate (**140**) with nitric acid affords the nitro derivative. Hydrogenation converts this to the anthranilate (**141**). In one of the standard conditions for forming quinazolones, that intermediate is then treated with ammonium formate to yield the hetero-cycle (**142**). Reaction of **142** with phosphorus oxychloride leads to the cor-responding enol chloride (**143**). Condensation of **143** with *m*-iodoaniline (**144**) leads to displacement of chlorine and consequent formation of the aminoquinazoline (**145**). Reaction with the trimethylsilyl derivative of acetylene in the presence of tetrakis-triphenylphosphine palladium leads to replacement of iodine by the acetylide. Tributylammonium fluoride then removes the silyl protecting group to afford the kinase inhibitor **erlotinib** (**146**).[22]

The nucleophile-accepting acrylamide group in the kinase inhibitor **canertinib** (**154**) may lead to covalent binding of that group to sites on the enzyme and thus irreversible inhibition. The starting quinazolone (**147**) is available by some scheme such as that above. The carbonyl group is first converted to its enol chloride (**148**) by means of phosphorus oxychloride.

Displacement of this halogen by the amino group of the substituted aniline (**149**) then affords intermediate **150**. The labile fluoride on the quinazoline benzene is next displaced with the alkoxide from the morpholylpropyl alcohol (**151**) to afford the ether **152**. Catalytic hydrogenation serves to reduce the nitro group to the corresponding amine (**153**). Acylation of the newly formed amine with acryloyl chloride completes the synthesis of **154**.[23]

The preparation of yet another variant on the theme starts with quinazolone (**155**). Treatmment with methanesulfonic acid selectively cleaves the ether at the more electron-rich position to give the phenol (**156**). This functional group is then acylated by means of acetic anhydride (**157**). The ring carbonyl group is next converted to the enol chloride (**158**), in this case by means of thionyl chloride. Condensation with the same aniline used in the previous example leads to the secondary amine (**159**). The product is then saponified so as to remove the acetyl group (**160**). Alkylation of the enolate from treatment of the phenol with base with the chloropropyl morpholine (**161**) then affords **gefitinib (162)**.[24]

A somewhat different strategy is used to prepare **vandetanib (171)**. The starting material in this case is analogous to **155** above with the ether and free phenol ring oxygen atoms reversed. The ring nitrogen in this molecule is protected as its pivaloyloxymethyl (POM) derivative, a group that is stable to most conditions, but can be selectively cleaved by ammonia. Mitsonobu coupling of the free phenol **163** with the alcohol oxygen on hydroxymethyl piperidine (**164**) leads to the ether (**165**). The *t*-BOC protecting group on piperidine nitrogen is then removed with mild base **166**. Reductive alkylation of the newly revealed amine with formaldehyde and borohydride then affords the N-methylated derivative (**167**). The POM protecting group is then cleaved with ammonia (**168**). The quinazolone carbonyl is next converted to the enol chloride with thionyl chloride (**169**).

Displacement of that halogen by the amine on the aniline (**170**) adds the final ring affording **171**.[25]

163; 164; 165; R = tBuOCO; 166; R = H; 167; 168; 169; 170; 171

The structures of two recent quinazoline kinase inhibitors feature a somewhat different substitution pattern from the preceding compounds. **Lapatinib** (**182**), for example, features single side chain on the fused benzene ring; that group is, however, more complex than those in the earlier examples. Preparation of the ring that will attach at the 1 position

172; 173; 174; R = O; 175; R = H; 176; 177; 179; 178; 180

of the quinazoline involves first alkylation of the nitrophenol (**172**) with the benzyl bromide (**173**) to afford the ether (**174**). The nitro group is next reduced to the corresponding aniline. In a convergent sequence the quina- zolone (**177**) is then converted to its enol chloride (**176**). Reaction of the aniline (**175**) with the enol chloride leads to displacement of halogen and formation of **178**. Suzuki coupling of this product with the furan boric acid (**179**) in the presence of the usual palladium catalyst attaches the furyl aldehyde group to the fused benzene ring (**180**).

The aldehyde group on the newly attached furan ring is then used to further elaborate the side chain. Thus reductive amination of the carbonyl group with ammonia leads to primary amine **181**. This newly introduced function is next alkylated with 2-methylsulfonylethyl chloride. There is thus obtained the kinase inhibitor **182**.[26]

The preceding compounds all featured a more or less complex aromatic amine attached to the heterocyclic part of the quinazoline. Aliphatic nitro- gen by way of contrast occupies that position in the inhibitor **tandutinib** (**189**). Alkylation of the phenolic function in **183** with 3-chloropropyl tosylate affords the ether (**184**) from displacement of the tosyl group. Reaction with nitric acid gives the ortho nitro derivative (**185**). Catalytic hydrogenation then reduces this to the corresponding amine. Treatment of this intermediate with formamide then adds the requisite atoms for forming the quinazoline ring. The carbonyl group is then converted to the enol chloride (**187**) by means of thionyl chloride. The sequence departs from previous schemes by the use of an alycyclic amine in the next step. Thus, the reactive enol halogen atom is displaced by the free amine in the monoacylated piperazine to afford **188**. Reaction of the product with piperidine under somewhat more forcing conditions with replaces the terminal chlorine on the ether-linked side chain to complete the synthesis of **189**.[27]

C. Miscellaneous Benzo-Fused Heterocycles

Aldose reductase inhibitors are expected to protect long-term diabetic patients from the consquences of the accumulation of sorbitol that is one of the consequences of the disease [see lidorestat (Chapter 7)]. A quinazolodione provides the nucleus for another potential drug in this class. The sequence for the preparation of this agent starts with the heterocycle (**190**), which is in essence simply the cyclic carbonate of the corresponding anthranilic acid. Heating the compound with the substituted benzylamine (**191**). Result in formation of the ring-opened amide with loss of carbon dioxide **192**. The ring is close again this time as a quinolodione (**193**); the requisite carbonyl carbon is restored by means of carbonyl diimidazole. The anion from reaction of this last intermediate with sodium hydride is then alkylated with ethyl bromoacetate. Saponification of the ester completes the preparation of **zenarestat** (**194**).[28]

Fluid accumulation is one of the graver consequence of congestive heart failure as this excess blood volume places additional strain on the weakening heart. The hormone vasopressin also known as antidiuretic hormone contributes to this effect. Though diuretics are often used to decrease this the excess fluid these drugs often upset the balance of electrolytes in the remaining fluid can thus adversely affect kidney function. Thiazides are, for example, well known to cause excretion of potassium. A recently developed non-peptide vasopressin antagonist has shown promising initial activity in relieving heart failure associated fluid retention without an effect on electrolyte balance. Construction of the benzapine moiety begins with esterification of anthranilic acid (**195**) followed by reduction of the nitro group with stannous chloride (**196**). The aniline nitrogen is then converted to *p*-toluenesulfonamide (**197**). Reaction of **197** with ethyl ω-chlorobutyrate in the presence of potassium carbonate then gives the alkylation product **198**. Potassium *tert*-butoxide-catalyzed Claisen condensation of this diester leads to azepinone **200** as a mixture of methyl and ethyl esters resulting from alternate cyclization routes. Strong acid leads to the transient ketoacid, which the decarboxylates; the toluenesulfonyl group is lost under reaction conditions to afford the azepinone (**201**). This last intermediate is then acylated with the benzoyl chloride **202** to afford amide **203**. Catalytic reduction of the nitro group proceeds to the aniline (**204**). The chain is next extended by acylation of the newly formed amine with *o*-toluyl choride (**205**) to give **206**. Reduction of the azepinone carbonyl group with borohydride affords the vasopressin antagonist **tolvaptan (207)**.[29]

REFERENCES

1. F. Labrie, Y. Merand, S. Gauthier, U.S. Patent 6,060,503 (2000).

2. V.A. Ashwood, R.E. Buckingham, F. Cassidy, J.M. Evans, E.A. Faruk, T.C. Hamilton, D.J. Nash, G. Stemp, K. Willcox, *J. Med. Chem.* **29**, 2194 (1986).

3. W.N. Chan, H.K.A. Morgan, M. Thompson, M. Evans, U.S. Patent 5,760,074 (1998).

4. S.L. Kattiger, R.G. Naik, A.D. Lakdawalla, A.N. Dohadwalla, R.H. Rupp, N.J. de Souza, U.S. Patent 4,900,727 (1990).

5. R.G. Naik, B. Lal, R.H. Rupp, H.H. Sedlack, G. Dickneite, U.S. Patent 5,284,856 (1994).

6. R.E. TenBrink, M.D. Ennis, R.A. Lahti, U.S. Patent 5,877,317 (1999).

7. M. Roe et al., *Bioorg. Med. Chem. Lett.* **9**, 595 (1999).

8. C.J. Torrance, P.E. Jackson, E. Montgomery, A. Wissner, K.W. Kinzler, B. Vogelstein, P. Frost, C.M. Discafani, *Nat. Med.* **6**, 1024 (2000).

9. S. Evangelista, *Curr. Opin. Inevest. Drugs* **6**, 717 (2005).

10. G.A.M. Giardina et al., *J. Med. Chem.* **42**, 1053 (1999).

11. D.B. Damon, R.W. Dugger, U.S. Patent 6,313,142 (2001).

12. N. Mealy, J. Castaner, *Drugs Future* **24**, 871 (1999).

13. M.G. Venet, P.R. Angibaud, P. Muller, G.C. Sanz, U.S. Patent 6,037,350 (2000).

14. P. Angibaud et al., *Bioorg. Med. Chem. Lett.* **13**, 1543 (2003).

15. T. Fujioka, S. Teramoto, T. Mori, T. Hosokawa, T. Sumida, M. Tominaga, Y. Yabuuchi, *J. Med. Chem.* **35**, 3607 (1992).

16. K. Masuzawa, S. Suzue, K. Hirai, T. Ishizaki, U.S. Patent 4,980,470 (1990).

17. J. Prous, J. Graul, J. Castaner, *Drugs Future* **19**, 827 (1994).

18. A. Graul, X. Rabasseda, J. Castaner, *Drugs Future* **24**, 1324 (1999).

19. A.S. Thompson, E.G. Corley, M.F. Huntington, E.J.J. Grabowski, *Tetrahedron Lett.* **36**, 8937 (1995).

20. A. Fensome et al., *J. Med. Chem.* **48**, 5092 (2005).

21. D.M. Manoury, J.L. Binet, A.P.D. Dumas, F. Lefevre-Borg, I. Cavero, *J. Med. Chem.* **29**, 19 (1986).

22. R.C. Schnuur, L.D. Arnold, U.S. Patent 5,747,498 (1998).

23. J.B. Small et al., *J. Med. Chem.* **43**, 1380 (2000).

24. A.J. Barker et al., *Bioorg. Med. Chem. Lett.* **11**, 1911 (2001).

25. L.F. Henenquin et al., *J. Med. Chem.* **45**, 1300 (2002).

26. K.G. Petrov et al., *Bioorg. Med. Chem. Lett.* **16**, 4686, (2006).

27. A. Pandey et al., *J. Med. Chem.* **45**, 3772 (2002).

28. M. Hashimoto, T. Oku, Y. Ito, T. Namiki, K. Sawada, C. Kasahara, Y. Baba, U.S. Patent 4,734,419 (1988).

29. K. Kondo et al., *Bioor. Med. Chem.* **7**, 1743 (1999).

CHAPTER 9

BICYCLIC FUSED HETEROCYCLES

1. COMPOUNDS WITH FIVE-MEMBERED RINGS FUSED TO SIX-MEMBERED RINGS

A. Compounds with Two Heteroatoms

The platelet aggregation inhibitors ticlopidine and clopidogrel consist of substitued thienopiperidines, as does the recent entry **prasugrel** (**9**). Alkylation of the enolate from treatment of *N*-(4-chlorobenzyl)piperidine, (**1**), with sodium hydride with ethyl chloroacetate affords the ketoester (**2**). Reaction of **2** with hydrogen sulfide leads to formation the thiolactone (**3**).[1] This transform can be rationalized by assuming initial conversion of the ketone to a thioenol; the thioenol attacks the ester to close the ring. Reaction of the product would then form the corresponding enol acetate (**4**). In a convergent sequence, the Grignard reagent from the benzyl bromide (**5**) is added to cyanocyclopropane (**6**). This afford ketone **7** after hydrolysis. Treatment of **7** with NBS leads to formation of the brominated derivative (**8**). Reaction of this last with the thienopiperidine (**4**) leads to displacement of bromine and formation of the alkylation product. Thus **9** is obtained.[2]

The Organic Chemistry of Drug Synthesis, Volume 7. By Daniel Lednicer
Copyright © 2008 John Wiley & Sons, Inc.

γ-Aminobutyric acid is the main inhibitory brain neurotransmitter and is thus involved in many psychological process. Muscimol (**10**), the GABA agonist from the *amanita* mushroom, for example, produces inebriation. A close analogue, **gaboxadrol** (**15**) in which the pendant side-chain amine is closed to form a piperidine is currently being investigated as a sleep aid. The drug reduces insomnia and improves the quality of sleep. The synthesis of this compound begins by forming the acetal from **11** by reaction of the compound with ethylene glycol. (**12**). Reaction of this intermediate with hydroxylamine replaced the ethoxy group in the ester to addord the hydroxamic acid (**13**). Treatment of that intermediate with acid leads first to hydrolysis of the ketal; reaction of the terminal hydroxyl on the hydroxylamine group with the ketone leads to closure to the isoxazole ring. Hydrolysis under more strenuous conditions frees the piperidine nitrogen to afford **15**.[3]

The ACE inhibitors, first introduced almost three decades ago, expanded the means of treating hypertension by providing yet another mechanism for lowering blood pressure. The pioneer drug, captopril, was followed on the market by well over a dozen other ACE inhibitors. Current research focuses on compounds that will inhibit vasopeptidases, a category of endogenous substances that includes not only ACE but a number of other enzymes involved in regulation of the cardiovascular function. The construction of the bicyclic nucleus for one of these agents starts by forming the amide (**18**) between trifluoroacetamide-protected cysteine (**16**) with the dimethylacetal (**17**). Treatment of that disulfide intermediate (**18**) with tributylphosphine leads to reduction of the disulfide bond to a thiol in effect cutting that precursor in two. That transient intermediate **19** is then immediately exposed to acid. The sequence that follows can be rationalized by assuming the thiol exchanges with one of the methoxyl groups on the acetal function to form a seven-membered ring; the remaining acetal methoxy is then replaced by the amide nitrogen to close the six-membered ring. Note that the carbon atom that results from this transform (**20**) has not changed oxidation state as it consists of the sulfur equivalent of a carbinolamine. Saponification of the trifluoroacetyl amide removes that protecting group to reveal the primary amine (**21**). This function is then acylated with the 2-thioacetyl phenylacetic acid (**22**). Removal of the protecting groups on this last intermediate affords the vasopeptidase inhibitor **omapatrilat (24)**.[4]

B. Compounds with Three Heteroatoms

Sleep inducing drugs go back to the very earliest days of medicinal chemistry. The earliest drugs comprise the barbiturates, which though effective had some serious limitations. The benzodiazepines that came along in the 1960s were better tolerated though these too had their drawbacks, such as drug hangover and the propensity to induce dependence. These have now been largely replaced by drugs, such as zolpidem. The structures of these drugs share the purine 6–5 fused rings system found in purines; the placement of ring nitrogens atoms is usually quite different from that found in purines. The synthesis of a recent example begins with the condensation of the substituted acetophenone (**25**) with DMF acetal. This reaction affords the chain-extended enamide (**26**). The amide nitrogen is then alkylated (**27**). Reaction of this intermediate with the aminopyrrazole (**28**) leads to formation of the fused pyrimidinopyrrazole (**29**). For bookkeeping purposes, the transform may be visualized to involve addition–elimination of the enamide diethylamino group by the amine on the pyrrazole. Imine formation between the carbonyl and pyrrazole ring nitrogen then closes the fused ring. The product (**29**) is next acylated with thiazole-carboxylic acid (**30**) in the presence of aluminum chloride (**31**). Thus, **indiplon (31)** is obtained.[5]

The enzyme, purine nucleoside phosphorylase (PNP), is directly involved with blood levels of T-cells. Low levels of this enzyme will inhibit T-cell proliferation. Drugs that inhibit the enzyme can also be expected to act against proliferation of malignant T-cells. The PNP inhibitor **forodesine (36)** has shown early activity against T-cell malignancies. Treatment of the deazapurine (**32**) with lithium leads to derivative **33** lithiated on the pyrrole ring. Condensation of that species with the highly protected aminosugar (**34**) results in addition of the anion to the imine function on the sugar. Deprotection of that product (**35**) in several steps then affords **36**.[6]

The pteridine, folic acid, comprises one of the essential factors required for synthesis of DNA and by extension cell proliferation. As a result, folate antagonists have been intensively investigated as potential antitumor compounds. The pteridine-based antagonist methotrexate, which was developed many years ago, simply comprises folic acid in which the side-chain amine is methylated and nitrogen replaces the oxygen atom at the 4 position. This drug is widely used as an antitumor compound. Methotraxate in fact comprises the "M" in the acronym of the many multidrug cocktails used by oncologists. Most of the more recent folate antagonist, as exemplified by palitrexate (**202**) and pelitrexol (**213**) described later in this chapter, retain the two fused six-membered rings found in folic acid and replace the side-chain amine group by a methylene group. Contraction of one of the rings, is interestingly consistent with folate antagonism. The first few steps in the construction of the nucleus of **pemetrexed** (**46**) follow a well-precedented scheme. Reaction of the enolate from ethyl cyanoacetate (**37**) with the methyl acetal from bromoacetaldehyde (**38**) leads to the alkylation product (**39**). Condensation of **39** with guanidine then forms the pyrimidine (**40**). The transient side-chain aldehyde from treatment of the compounds with acid then forms an imine with the adjacent amino group, thus closing the fused five-membered ring (**41**). The remaining primary amino group is then protected as its *tert*-butyl carbamate (**42**). Reaction of **42** with iodo sucinimide proceeds to form the 3-iodo derivative (**43**).

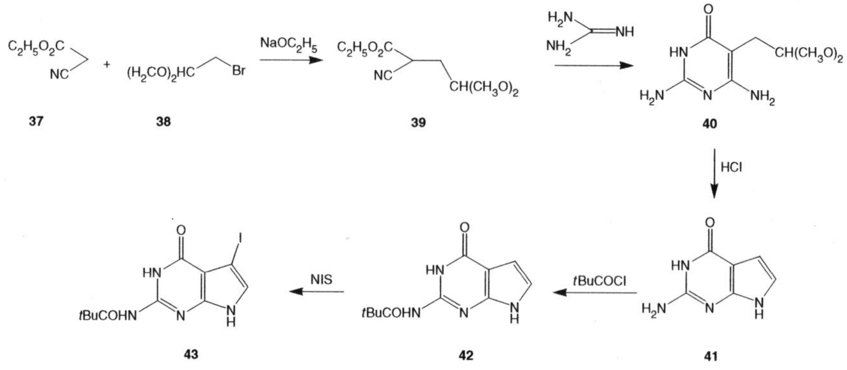

NIS = *N*-iodosuccinimide

The key step in this synthesis comprises grafting the benzenzamido-glutamate moiety found in folic acid onto the heterocycle. Thus, condensation of the iodo-substituted (**43**) with acetylide (**44**) catalyzed by tetrakis-triphenylphosphine palladium affords the coupling product (**45**). The triple bond is then converted to the saturated bridge by catalytic hydrogenation. Saponification then removes the protecting group and at the same time hydrolyzes the esters on the glutamate fragment to afford **46**.[7]

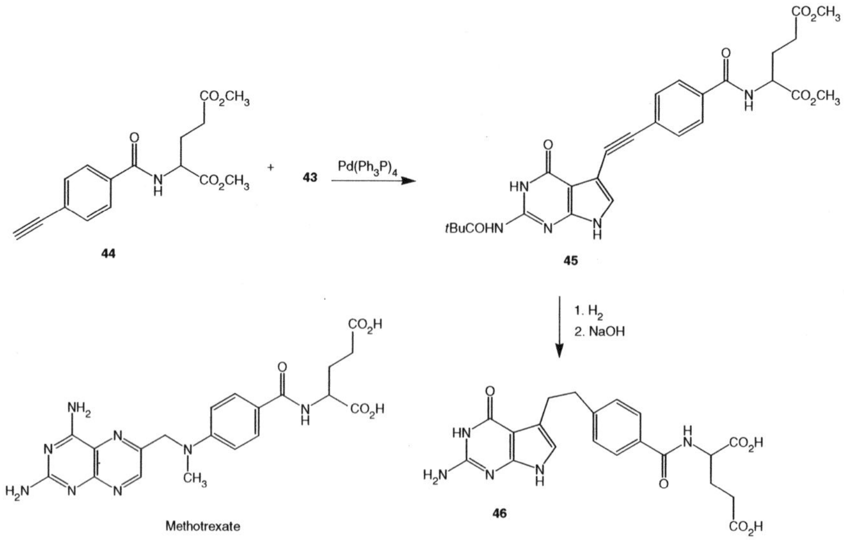

Methotrexate

C. Compounds with Four Heteroatoms

The A_1 adenosine receptors fulfill a largely inhibitory role. Research has thus recently focused on agonist with structures based on adenosine itself as agents that will overcome responses due to inappropriate excitation, such as tachycardia and some arrhythmias. Replacement of one of the hydrogen atoms on the exocyclic amine in adenosine by a tetrahydrofuryl group provides an effective A_1 adenosine agonist. Preparation of this fragment as a single enantiomer starts with a modern version of the Curtius reaction. Thus, reaction of tetrahydrofuroic acid (**47**) with triphenylphosphoryl azide leads to isocyanate (**48**). Treatment of this intermediate with benzyl alcohol then affords the corresponding carbamate (**49**). Catalytic hydrogenation removes the benzyloxy group leading to the free primary amine **50**. The product is then resolved by way of its camphorsulfonyl salt to afford **51**. Reaction of this intermediate with desamino chloroadenosine (**52**) affords **tecadenoson (53)**.[8]

Another approach to preparing A_1 adenosine receptors agonists involves converting the hydroxymethyl group on the sugar moiety to an amide in addition to adding a substituent to the amine of adenosine. The starting material (**56**) is arguably obtainable by oxidation of inosine acetonide (**54**), followed by acetylation of the hydrolysis product. Reaction of the acid (**55**) with thionyl chloride followed by ethanol affords the corresponding ethyl ester (**56**). The ring oxygen on this intermediate is next replaced by chlorine by means of phosphorus oxychloride to yield **57**. Reaction of **57** with cyclopentylamine displaces the halogen to form the cyclopentylamino derivative (**58**). Treatment with triethylamine under somewhat more strenuous conditions effects ester–amide interchange to form the amide; the acetyl protecting groups are cleaved under those reaction conditions. Thus, the A_1 adenosine receptors agonist **selodenoson** is obtained (**59**).[9]

Yet another modification leading to an adenosine agonist involves conversion of one of the amino groups on the fused pyrimidine ring of adenosine to a pyrazole. The synthesis begins with the conversion of guanosine to its 5′ acetate by reaction with acetic anhydride. The hydroxyl at the 4 position is replaced by chlorine in the usual manner by treatment with phosphorus oxychloride to afford **61**. In a variation on the Sandmeyer reaction, this last intermediate is allowed to react with amyl nitrite in the presence of methylene iodide in an aprotic solvent. The low concentration of nitrite ion from decomposition of its amyll ester serves to diazotize the amine; the diazonium intermediate then captures iodine from the other reagent to form the iodo derivative (**62**). Reaction of **62** with ammonia interestingly proceeds selectively at the 4 position on the purine to afford **63**. Treatment of this product with hydrazine, presumably under more strenuous conditions, displaces iodine to form **64**. Condensation of **64** with carbethoxymalonaldehyde (**65**) then affords the pyrazole ring, the product from formation of an imine from each aldehyde (**66**). Replacing the ester with an amide by ester interchange with methylamine completes the synthesis of **regadenoson** (**67**).[10]

One of the most fruitful approaches for designing antiviral compounds comprises administration of false substrates that will halt replication of the viruses. Such agents will bring replication to a stop in the event they are mistaken for the real thing. This strategy was validated some decades ago by the antiviral agent acyclovir in which the sugar moiety is replaced by an open-chain surrogate. Phosphorylation comprises one of the first steps in incorporation of a nucleoside into DNA or RNA. Replacing the phosphate by an analogous function that is mistakenly taken up by the virus, but cannot be further processed, comprises yet another strategy for halting viral replication. The HIV reverse transcriptase inhibitor **tenofovir** (**78**), combines both those strategies in a single molecule. As a first step, hydroxy ester **68** is protected by reaction with THP to form the corresponding ether (**69**). The ester function is the reduced to the alcohol (**70**) by means of Selectride. The newly formed alcohol function is then activated toward displacement by conversion to its tosylate (**71**). Reaction of **71** with adenine (**72**) attaches the much abbreviated side chain to the five-membered ring (**73**). The exocyclic amine function is then protected as its benzamide (**74**) by means of benzoyl chloride and the THP group removed with aqueous acid. The alkoxide from reaction of this last intermediate (**75**) with sodium hydride is next alkylated with the tosyl derivative of methyleneisopropyl phosphate (**76**) to afford the methylene phosphoryl product (**77**). Removal of the protecting groups completes the synthesis (**78**).[11]

Replacement of one of the carbon atoms in the sugar moiety by oxygen, in effect converting that ring to an acetal, leads to yet another false substrate for viral reverse transcriptase. Glycosilation of the silylated purine (**79**) with chiral dioxolane (**80**), prepared in several steps from anhydromannose, in the presence of ammonium nitrate affords the coupling product as a mixture of anomers. The mixture of products is then separated on a chromatographic column. The desired diastereomer (**81**) is the reacted with ammonia to afford the product (**82**) from replacement of fluorine. Reaction of this intermediate with ammonia under more strenuous conditions replaces chlorine (**83**). Treatment **83** with fluoride ion removes the silyl protecting group to afford the reverse transcriptase inhibitor **amdoxovir** (**84**).[12]

The surrogate sugar moiety in the antiviral **omaciclovir** (**90**) like that in acyclovir, consists of an open-chain alcohol. The presence of two hydroxyl groups more closely mimics the structure of a sugar than does the side chain in the latter. This agent shows good activity against varicela zoster, the cause of chicken pox and shingles. The scheme for preparing this compound differs from the preceding example in that the pendant group is attached via a Michael reaction. Thus reaction of the purine (**85**) with exomethylene glutarate (**86**) in the presence of base leads to conjugate addition of purine nitrogen to the double bond (**87**). The carboxylates are next reduced to the corresponding alcohols by means of borohydride to afford **88** as a mixture of enatiomers. Reaction with ammonia then replaces chlorine on the purine by nitrogen to afford the diamine (**89**). This intermediate is then treated with the enzyme adenosine deaminase immobilized on a solid support. The enzyme preferentially reacts with the isomer in which the stereochemistry of the secondary hydroxyl group more closely resembles that in the furanoside in a natural nucleoside converting the amine at the 4 position to a hydroxyl group. Separation of that from unreacted starting material affords **90**.[13]

The concept of inhibiting cell growth by providing false substrates in fact well predates its application to antiviral agents. The first HIV reverse transcriptase drug, zidovudine (AZT) was actually first prepared as a potential anticancer agent. An enzymatic reaction provides a one-step entry to the antileukemia compound **nelarabine** (**94**). Thus reaction of 6-methoxyguanine (**91**), with uracyl arabinoside (**92**) in the presence of the enzymes uridine phosphorylase and purine nucleotide phosphorylase in effect transfer the sugar from the pyrimidine to the purine, yielding **93**.[14]

The antitumor chemotherapy agent **clofarabine** (**104**) sports halogen on the purine as well as on the sugar. The major portion of the synthesis involves preparation of the requisite fluorine-substituted furanose (**101**). The sequence begins with the conversion of the only free hydroxyl in allofuranose bis(acetonide) (**95**) to its tosylate (**96**). This group is then displaced by fluorine in its reaction with potassium fluoride to afford (**97**). Acid hydrolysis then removes the two acetonide protecting groups. Reaction with benzoyl chloride proceeds to form the benzoate ester (**98**) from the most sterically accessible hydroxyl group. Treatment of **98** with periodate cleaves the 2,3 vicinal diol to form a transient inter-mediate, such as **99**. What had been formerly one of the side-chain hydroxyl groups then forms an acetal with one of the newly revealed alde-hydes to afford the new furanose (**100**) (the product from attack on the other carbonyl would be a dioxolane, which would reopen under reaction conditions). Reaction with acetic anhydride converts the anomeric hydroxyl to its acetate. That group is then replaced by the better leaving group bromine by means of hydrogen bromide (**101**). Glycosylation of the trimethylsilylated purine (**102**) leads to the coupling product (**103**). Ammonia then replaces chlorine at 4 by an amine and at the same time removes the benzoyl protecting group. Thus, clofarabine (**104**) is obtained.[15]

Nerve cells are known to elaborate proteins that guide early development of the nervous system and in later life play a role in cell repair and regeneration. A purine that crosses the blood–brain barrier has shown activity similar to those endogenous factors *in vitro* as in *in vivo* models of nerve damage. Conjugate addition of adenine to ethyl acrylate in the presence of base leads to the ester (**106**). Reaction of that product with sodium nitrite in the presence of acid converts the amino group at the 4 position to the corresponding hydroxyl derivative. Saponification then leads to the free carboxylic acid, shown as its keto tautomer (**107**). The acid is next esterified with trifluoroacetyl nitrophenol (**108**) in order to provide that function with a good leaving group (**109**). Ester–amide interchange with ethyl *p*-benzoate (**110**) displaces nitrophenol to form the corresponding benzamide. Base hydrolysis of the terminal ester then affords **leteprinim (111)**.[16]

The xanthine theophylline has been used for well over a century for treating asthmatic episodes by relaxing the constricted bronchioles that are part of attacks. The discovery of the role of the phosphodiesterase enzyme (PDE) by Sutherland in the early 1960s helped explain the mode of action of this drug. Research over the past few decades has led to the identification of a large number of subtypes of PDE receptors. Theophylline, for one, was found to interact mainly with PDE IV receptors. The relatively narrow therapeutic index of this drug has led to the continued search for better tolerated agents. The synthesis of a recent example starts with nitrosation of the pyrimidinedione (**112**) with sodium nitrite in the presence of acid to yield **113**. Reduction of the newly introduced nitroso group with sodium dithionite leads to the 1,2-diamine (**114**). Reaction of **114** with formamide closes the fused imidazole ring. Thus, the PDE IV inhibitor **arofylline (115)** is obtained.[17]

Replacing the hydrogen atom on the imidazole ring of a purine with a large lipophillic group effects a major change in the biological activity. The resulting compounds are no longer PDE inhibitors, but instead act as adenosine receptor antagonists. The first of these, **istradefylline (120)**, selectively blocks A(2A) receptors and in addition inhibit monoamine oxidase B (MAO-B). The neuroprotective action of this agent in animal models for Parkinson's disease has been attributed to this mixed activity. Reaction of the diamine (**116**), with the dimethoxy cinnamic acid (**117**) in the presence of a diimide leads to the formation of a mixture of the amides formed between the acid and one of the two amines on the pyrimidine (only one, **118**, is shown). Heating that mixture with sodium hydroxide leads to cyclization, forming the xanthine (**119**). The free amine on the fused imidazole is then alkylated with methyl iodide in the presence of base to afford **120**.[18]

EDAC = 1-ethyl-3-(3-dimethylaminopropyl) carbodimide

Substitution by a somewhat complex bridged biyclic moiety provides a compound that acts as an adenosine A_1 receptor antagonist. These ubiquitous receptors, which are involved in regulating oxygen consumption and the flow of blood in cardiac tissue, generally suppress those functions. An antagonist would thus be potentially useful for treatment of congestive heart failure. The xanthine portion of this molecule is constructed much as that in the previous case. Thus, acylation of the diamino pyrimidine (**121**) with acryloyl chloride in this case affords a single amide (**123**). Reaction with base then closes the imidazole ring. Diels–Alder condensation of the vinyl group on this ring with cyclopentadiene proceeds to form the bridged bicyclic system (**125**). Oxidation of the isolated double bond on that newly formed moiety with (MCPBA) then affords the oxirane (**126**). Thus the adenosine antagonist **naxifylline** is obtained.[19]

Adding yet another large substituent to the imidazole ring restores PDE inhibiting activity. The resulting compound **dasantafil (137)** blocks PDE 5 and thus joins the growing list of agents that address erectile disfunction. The synthesis of this compound begins with the formation of the reductive amination product **(129)** from anisaldehyde **(128)** and glycine. This product is then treated with reagent **130**, obtained from reaction of triethyl orthoformate and cyanamide in the presence of strong base. For bookkeeping purposes formation of the imidazole **(131)** can be rationalized by assuming initial addition–elimination of the amine in **129** to the reagent **130**; addition of the enolate from **129** to the nitrile then closes the ring. Condensation of the product **(131)** from this reaction with diethylcarbamate again in the presence of strong base forms the pyrimidinedione ring to afford the xanthine **(132)** after neutralization. The nitrogen at position 1 is next alkylated with bromoethyl acetate **(133)**. Reaction of **133**

with NBS results in bromination on both the aromatic ring and the carbon on the fused imidazole ring **(134)**. Displacement of the latter bromine atom with chiral aminocyclopentanol **(135)** completes construction of the skeleton (**136**). Saponification of the acetate group on the pendant side chain then affords the PDE V inhibitor **137**.[20]

Moving one of the imidazole nitrogen atoms found in xanthines into the ring junction provides the nucleus for another PDE 5 inhibitor. The other substituents in this compound more closely resemble those found in sildenafil (aka Viagra) than those used in dasantafil. The convergent scheme starts with the acylation of alanine (**138**) with butyryl chloride. The thus-produced amide (**140**) is again acylated, this time with the half acid chloride from ethyl oxalate in the presence of DMAP and pyridine to afford intermediate **141**. In the second arm of the scheme, the benzonitrile (**142**) is reacted with aluminate **143**, itself prepared from trimethylaluminum and ammonium chloride, to form the imidate (**144**). Treatment of **144** with hydrazine leads to replacement of one of the imidate nitrogen atoms by the reagent in an addition–elimination sequence to form **145**. Condensation of **145** with **141** leads to formation of the triazine (**146**). Phosphorus oxychloride then closes the second ring to afford **147**. Reaction of this last intermediate with chlorosufonic acid affords the corresponding sulfonyl chloride. Treatment of this compound with *N*-ethyl piperazine forms the sulfonamide and thus **vardenafil (148)** is obtained.[21,22]

The peptide hormone, glucagon-like peptide (GLP-1), which is released in response to intake of food, regulates insulin levels. The hormone, in the normal course of events, exerts only momentary activity as it is quickly inactivated after its release by a dipeptidal peptidase (DPP). Inhibitors of that degrading enzyme should thus lead to persistent levels of GLP-1. The search for small molecule inhibitors of DPP that prolong the action of GLP-1 has consequently been the focus of research for finding alternate methods for controlling Type 2 diabetes. Construction of the heterocyclic nucleus (**153**) begins with the displacement of chlorine in pyrazine (**149**) with hydrazine to afford **150**. Acylation with trifluoroacetic anhydride proceeds on the more basic terminal nitrogen to afford the trifluoroacetohydrazide (**151**). Treatment of this intermediate with polyphosphoric acid leads the enol form of the amide to undergo addition–elimination with the pyrazine nitrogen and to form the new fused ring (**152**). Catalytic hydrogenation proceeds to selectively reduce the six-membered ring to afford the triazolopyrazine (**153**).

In the other arm of the convergent scheme, the bis(lactam) enol ether (**154**) can be viewed as a latent chiral carboxyl group since the transannular isopropyl group will transfer its chirality across the ring to a new entering moiety. Thus, alkylation of the enolate from **154** with the trifluorobenzyl chloride (**155**) yields (**156**) as virtually a single enantiomer. Methanolysis of the product followed by bis-*t*-BOC affords the corresponding *t*-BOC protected amino acid. Saponification then yields the carboxylic acid **157**. This product is next subjected to Arndt–Eistert homologation. The carboxylate is first converted to its acid chloride with *tert*-butyl chloroformate. This intermediate affords diazoketone **158** with diazomethane. Treatment with silver benzoate causes this reactive species to rearrange to the homologated ester (**159**). The acid (**160**) obtained on saponification is then coupled in the presence of a diimide with the heterocyclic nucleus **153** to afford the amine **161**. Removal of the *t*-BOC protecting group from **105** with strong acid completes the synthesis of **sitagliptan (162)**.[23]

The majority of nucleoside antiviral agents, as noted previously, inhibit the production of new virions by acting as false substrates for the enzymes that synthesize viral DNA or RNA. The nucleoside-like compound, **isatoribine (170)** has shown activity against hepatitis C, a viral infection unusually refractory to treatment. This agent interestingly owes its efficacy to its stimulation of the immune system rather than to direct antiviral action. Condensation of the substituted malonate (**162**) (obtained by alkylation of 2-bromomalonate with 2-trimethylsilyl ethylmercaptan) with guanidine affords the pyrimidine (**163**). The function at the 4-position is then converted to the corresponding halide by reaction with phosphorus oxychloride to yield **164**. Addition–elimination of the pyrimidine hydroxyl with the aminated sugar (**165**) adds the requisite saccharide moiety to the compound **166**. The bridging amino group is next acylated with ethyl chloroformate to yield **167**. Treatment of **167** with tributylammonium fluoride leads to loss of the silyl group on the terminus of the ethyl mercaptide. The chain on sulfur then collapses to leave behind an anion on that atom. This ion then displaces the ethoxide from the carbamate on the proximate nitrogen in effect closing the thiazolone ring; the silyl protecting group on the sugar also cleaves under reaction conditions to yield the free carbinol (**168**). The chlorine atom at the 4 position, which has served to protect that function over the preceding steps, is then replaced with sodium methoxide to form the methyl ether (**169**). Trimethyl silyl iodide then cleaves that ether to leave behind a hydroxyl group. Acid hydrolysis then opens the acetonide to afford the immune system stimulant **170**.[24]

2. COMPOUNDS WITH TWO FUSED SIX-MEMBERED RINGS

Benzodiazepines provided the first relatively well-tolerated drugs for treating anxiety. These drugs were a major advance over the previously used barbiturates. A series of side effects, however, became evident over time as the benzodiazepine gained mass usage. The most common of these was habituation, a condition just short of true addiction. The more recent anxiolytic compounds of the "spirone" class introduced by buspirone are generally free of those limitations. A compound with a markedly different structure from these agents showed promising activity in animal models. The pyridine dicarboxylic acid (**171**) is first converted to its acid chloride with thionyl chloride; reaction with methanol then affords the ester **172**. Catalytic hydrogenation then serves to reduce the pyridine ring to a piperidine of undefined stereochemistry (**173**). Reaction of **173** with chloroacetonitrile affords **174**. Treatment of **174** with Raney nickel reduces the cyano group to the corresponding primary amine; this product then undergoes internal ester–amine interchange to yield the cyclized amide **175**. Lithium aluminum hydride then serves to reduce the amide to an amine; the ester on the other ring is converted to a carbinol in the process, affording the amino alcohol (**176**). The basic function is next alkylated with 2-chloropyrimidine (**177**). Reaction of the alcohol in **178** with methanesulfonyl chloride leads to the mesylate; this group is next displaced by means of sodium azide and the newly introduce azide group reduced to the primary amine. Resolution of this last product as its mandelate salt then yields **179** as a single enantiomer. Reaction of **179** with succinic anhydride converts the pendant amine to a succinimide affording the anxiolytic agent **sunepitron (180)**.[25]

The principal metabolite of one of the more recent antipsychotic drugs, risperidone, has significant dopamine blocking activity in its own right. As is often the case, where a metabolite has the same effect as the ingested drug, this opens the possibility that the metabolite is responsible for the observed effect. Condensation of acetoacetamide (**182**), arguably available in several steps from risperidone intermediate **181**,[26] with the cyclic imidate (**183**) affords the metabolite as its O-methyl ether **184**. Cleavage of that ether, for example, with boron tribromide, would then afford **paliperidone (185)**.[27]

The structure of a recent quinolone antibiotic differs from the majority of those agents in the nature of the substituent on the fused aromatic ring; the inclusion of nitrogen in that ring is unusual though not unprecedented. This structural feature in fact harks back to the very first quinolone, nalidixic acid, as well as the compound nofloxacin, which led to the revival of this field some two decades ago. Synthesis of the novel side-chain amine comprises the larger part of the preparation of the quinolone antibiotic **gemifloxacin (196)**. The sequence starts with conjugate addition of glycine ethyl ester **186** to acrylonitrile. Reaction of **187** with strong base leads to addition of the enolate adjacent to the nitrile to the ester carbonyl to afford the pyrrolidinone (**188**). The amine is then protected as its *tert*-butoxycarbamate (**189**) by reaction with bis-*t*-BOC. Treatment with sodium borohydride serves to reduce the ring carbonyl to the alcohol. The cyano group is next reduced to the corresponding primary amine by means of lithium aluminum hydride and this new function then also protected as its *tert*-butoxycarbamate (**191**). Oxydation of the ring alcohol with the sulfur trioxide: pyridine complex restores the ring carbonyl group (**192**). This function is then converted to its methoxime (**193**) by reaction with *O*-methylhydroxylamine. Treatment with acid then removes the *tert*-butoxycarbamate protecting groups to reveal the basic amino groups (**194**). Reaction of **194** with the often used quinolone starting material **195** results in displacement of chlorine by the more basic of the two amino groups in **194**. Thus, the antibiotic **196** is obtained.[28]

The search for better tolerated folate antagonist antitumor agents continues to produce yet more variants on methotrexate [see pemetrexed (**46**) for structure]. The analogue **pralatrexate (202)** retains most of the features of the prototype, but replaces the side-chain tertiary amine by a methylene–propargyl group. Alkylation of the enolate from dimethylhomotetrephthalate (**197**) with propargyl bromide attaches that group to the benzylic carbon to yield **198**. Reaction of the enolate from that compound with the bromomethylpteridine (**199**) yields the alkylation product **200**. Saponification of the esters in that product that leads to the free acid. The carboxyl group on the tertiary benzylic carbon decarboxylates on heating to yield the key intermediate **201**. This compound is then converted to a folate-like compound by reaction of the acid chloride with diethylglutamine. Saponification of diethylglutamine yields the free acid and thus **202**.[29]

A more recent example omits one of the nitrogen atoms in the pteridine ring and replaces the linking benzene ring of the side chain by thiophene. The convergent synthesis begins with the palladium-catalyzed coupling of the brominated pyridopyrimidone (**203**) with trimethylsilyl acetylene to afford the ethynyl derivative **203**. The silyl protecting group is then removed to afford **204**. In the second arm of the scheme, thiophenecarboxylic acid (**205**) is treated with bromine to afford the brominated derivative **206**. The carboxylic acid is then esterified with ethanol to yield **207**. Palladium-catalyzed coupling of the thiophene derivative (**207**) with the ethynyl derivative **204** then affords **208**. Catalytic hydrogenation reduces both the acetylene bond and the heterocyclic ring to which it is attached to afford **209** as a mixture of enatiomers.[30]

The fact that the final compound will contain chiral atoms at two remote locations on the fused piperidine ring and on the yet-to-be added glutamine moiety adds complications to the preparation of an enantiometrically pure product. The problem was solved by resorting to the use of enzymes. The first step comprises treatment of intermediate **209** with base to hydrolyze the ester. The secondary amine on the fused piperidine ring is then converted to an oxamate (**210**) by reaction with the half acid chloride of ethyl oxalate. The glutamate side chain is then added (**211**). Digestion of this intermediate with a lipase from *Candida antarctica* selectively hydroylizes the terminal ester on the glutamate of the (*S*) enantiomer allowing this compound to be easily separated from its epimers. Base hydrolysis of this compound then provides **pelitrexol (213)**.[31]

EDC = 1-ethyl-3-(3-dimethylaminopropyl) carbodeimide hydrochloride

3. MISCELLANEOUS COMPOUNDS WITH TWO FUSED HETEROCYCLIC RINGS: BETA LACTAMS

The overwhelming majority of semisynthetic beta-lactam antibiotics, the penicillins and cephalosporin, currently available to physicians trace their origins to the intense research effort devoted to this field several decades ago. The emergence of pathogens resistant to those antibiotics has led some laboratories to revisit this field. The modified cephalosporin **ceftobiprole (220)**, a compound with a rather complex extended side chain, has shown activity in the clinic against some strains of multidrug resistant bacteria. The synthesis starts with the well-precedented acylation of the of the cephalosporin (**215**), available in several steps from the commercially available 7-acetoxy cephalosporanic acid, with the activated thiadiazole carboxylic acid (**214**). The hydroxyl group in the product (**216**) is then oxidized with manganese dioxide to afford the corresponding aldehyde (**217**). This product is condensed with the bis(pyrrolidyl phosphonium) salt (**218**), itself protected with the complex carbonate (**219**). Removal of the several protecting groups then affords the highly modified cephalosporin **220**.[32]

Yet another recent beta-lactam active against multidrug resistant organisms is based on the "unnatural" carbapenem nucleus first used for the antibiotic imipenem. The preparation of the side chain starts with the protection of the nitrogen in hydroxyproline (**221**) as its diisopropyl phosphoryl derivative (**222**). The carboxyl is then activated as the mixed anhydride (**223**) by reaction with diphenylphosphinic anhydride. The ring hydroxyl is next converted to its mesylate (**224**) by reaction with methanesulfonyl chloride. Treatment of **224** with sodium sulfide serves to replace phosphorus on the carboxyl group by sulfur (**225**). Under somewhat basic reaction conditions, the sulfide anion slowly displaces the transannular mesylate group to form a bridged thiolactone ring. The stereochemistry at the new carbon sulfur bond is inverted in the process (**226**). This intermediate is then treated with 3-aminobenzoic acid. The thiolactone now opens to form amide **228** thus completing the side-chain. The requisite beta-lactam intermediate **229** is perhaps surprisingly available commercially. Reaction of that with the side-chain intermediate (**228**) leads to replacement of phosphorus by the side-chain thiol function. Removal of the protecting groups then affords the carbapenem **ertapenem** (**230**).[33]

REFERENCES

1. J.-P. Maffrand, N. Suzuki, K. Matsubayashi, S. Ashida, U.S. Patent 4,424,356 (1984).

2. J. Brandt, N.A. Farid, J.A. Jakubowski, K.J. Winters, World Patent 2004/098713 (2004).

3. P. Krogsgaard-Larsen, U.S. Patent 4,301,287 (1981).

4. J.L. Moniot et al., U.S. Patent 6,162,913.

5. R.S. Gross, K.M. Wilcoxen, R. Ogelsby, U.S. Patent 6,472,528.

6. G.B. Evans, R.H. Fourneaux, V.L. Schramm, V. Singh, P.C. Tyler, *J. Med. Chem.* **47**, 3275 (2004).

7. E.C. Taylor, D. Kuhnt, C. Shih, M. Rinzel, G.B. Grindey, J. Barredo, M. Jannatipour, R.G. Moran, *J. Med. Chem.* **35**, 4450 (1992).

8. R.T. Lum, J.R. Pfister, S.R. Schow, M.M. Wick, M.G. Nelson, G.F. Schreiner, U.S. Patent 5,789,416 (1998).

9. H.C. Williams, W.C. Patt, U.S. Patent 4,868,160 (1989).

10. J.A. Zabocki, E.O. Elzein, V.P. Palle, U.S. Patent 6,403,567.

11. L.A. Sobrera, J. Castaner, *Drugs Future* **23**, 1279 (1998).

12. R.F. Shinazi, U.S. Patent 5,444,063 (1995).

13. P. Engelhardt, M. Hogberg, N.-G. Johanssen, X.-X. Zhou, B. Lindborg, U.S. Patent 5,869,493 (1999).

14. R.N. Patel, *Curr. Opin. Drug Disc. Dev.* **9**, 744 (2006).

15. K. Chilman-Blair, N.E. Mealy, J. Castaner, *Drugs Future* **29**, 112 (2004).

16. A. Graul, J. Castaner, *Drugs Future* **22**, 945 (1997).

17. A.V. Averola, J.M.P. Soto, J.M. Mauri, R.W. Gristwood, U.S. Patent 5,223,504 (1993).

18. J.P. Petzer, S. Steyn, K.P. Castagnoli, J.-F. Chen, M.A. Schwartzschild, C.J. Van der Schyf, N. Castagnoli, *Bioor. Med. Chem.* **11**, 1299 (2003).

19. L. Belardinelli, R. Olsson, S. Baker, P.J. Scammells, U.S. Patent 5,466,046 (1995).

20. V.A. Dahanukar, H.A. Nguyen, C.A. Orr, F. Zhang, A. Zavialov, U.S. Patent 7,074,923 (2006).

21. H. Haning, U. Niewohner, T. Schlenke, M. Es-Sayed, G. Schmidt, T. Lampe, E. Bischoff, *Bioorg. Med. Chem. Lett.* **12**, 865 (2002).

22. For alternate methods, see P.J. Dunn, *Org. Proc. Res. Dev.* **9**, 88 (2005).

23. D. Kim et al., *J. Med. Chem.* **48**, 141 (2005).

24. M.G. Goodman, E.P. Gamson U.S. Patent 5,166,141 (1992).

25. G.N. Bright, K.A. Desai, U.S. Patent 5,122,525 (1992).

26. D. Lednicer, "The Organic Chemistry of Drug Synthesis", Volume 5, John Wiley & Sons, Inc., NY, 1995, p. 151.

27. C.G.M. Janssen, A.G. Knaeps, E.J. Kennis, J. Vandenberk, U.S. Patent 5,158,952 (1992).

28. C.Y. Hong, Y.K. Kim, J.H. Chang, S.H. Kim, H. Choi, D.H. Nam, Y.Z. Kim, J.H. Kwak, *J. Med. Chem.* **40**, 3584 (1997).

29. J.I. DeGraw, W.T. Colewell, J.R. Piper, F.M. Sirotnak, *J. Med. Chem.* **36**, 2228 (1993).

30. E.Z. Dovalsantos, J.E. Flahive, J.B. Halden, M.B. Mitchell, W.R.L. Notz, Q. Tian, S.A. O'Neil-Slawecki, World Patent, WO2004113337 (2004).

31. S. Hu, S. Kelly, S. Lee, J. Tao, E. Flahive, *Org. Lett.* **8**, 1653 (2006).

32. P. Hebeisen, H. Hilpert, R. Humm, U.S. Patent 6,504,025 (2003).

33. K.M. Brands et al., *J. Org. Chem.* **67**, 4771 (2002).

CHAPTER 10

POLYCYCLIC FUSED
HETEROCYCLES

A number of compounds defy ready assignment to one of the structural categories that have been used to organize the agents described so far in this volume. The structures of the agents in this chapter, except for of the two camptothecins, have little in common with each other beyond the fact that they do not fit previous categories. The bibliographic details on several compounds indicate that they were first prepared as much as three decades ago. They appear here because they were granted a nonproprietary name relatively recently.

1. COMPOUNDS WITH THREE FUSED RINGS

Some of the very earliest antitumor compounds incorporated in their structure a bis(chloroethylamino) moiety, a group often referred to as a nitrogen mustard. This quite reactive group kills cells by alkylating DNA, and in effect inactivating this genetic material by forming covalent cross-links between the bases. Alkylating activity is, however, not restricted to DNA as these highly reactive agents attack many other tissues. Drugs that include this moiety are, as a result, associated with a veritable host of very serious side effects. The indiscriminate killing action of alkylating

The Organic Chemistry of Drug Synthesis, Volume 7. By Daniel Lednicer
Copyright © 2008 John Wiley & Sons, Inc.

agents is not restricted to eukaryotic cells; the compounds also inactivate microbes. A recent nitrogen mustard-containing compound is proposed for use *ex vivo* to inactivate pathogens in blood transfusion supplies. The acridine portion of **amustaline** (**6**) is intended to intercalate in DNA, a well-known property of this structural element, and thus bring the mustard to the intended site of action. Excess drug that is not complexed can be expected to be destroyed by blood enzymes or even by simple hydrolysis. This will minimize exposure to unreacted mustard by a patient who received treated blood. Construction of this agent starts by displacement of chlorine from 9-chloroacrdine (**1**) by the terminal amine on a so-called β-alanine ester (**2**) in the presence of sodium methoxide. Saponification then yields the corresponding acid (**3**). Esterification of the carboxyl group with triethanolamine (**4**) leads to the ester (**5**). The free hydroxyl groups on this intermediate are then replaced by chlorine by reaction with thionyl chloride. Thus, **6** is obtained.[1]

Increased awareness and the rising incidence of Type 2 diabetes among the aging population has led to the search for alternate drugs to the traditional hypoglycemic agents for treating the disease. Agents, such as vidagliptin (Chapter 5) and sitagliptan (Chapter 9) represent antidiabetic agents for that act by mechanisms that differ significantly from the traditional drugs. A naphthothiophene moiety provides the nucleus for a drug that acts as an insulin sensitizer. This compound in addition acts as a peroxisome proliferator activated receptor agonist (PPAR). In broad terms this agent activates liver enzymes that metabolize fatty acids thus lowering serum triglyceride levels. Reaction of the lithio reagent from 2,3-dimethylthiophene (**7**) with benzaldehyde leads to the carbonyl addition product (**9**). The benzylic hydroxyl is next removed by hydrogenation over palladium to yield **10**. This intermediate is then treated with the substituted benzoyl chloride (**11**) in the presence of stannic chloride to afford the product from acylation on the carbon with highest electron density, thus yielding the thiophene (**12**). Exposure of **12** to boron tribromide in the cold leads to formation of a new ring; the methyl ether on the pendant benzene ring is cleaved to the corresponding phenol in the process (**13**). The enolate from treatment of that phenol with potassium carbonate with the α-bromo ester (**14**), affords the corresponding alkylation product as a mixture of enantiomers. Saponification of the ester affords corresponding carboxylic acid. Resolution as its salt completes the synthesis of **ertiprotafib** (**15**).[2]

A compound whose structure bears some slight resemblance to **15**, based on a phenoxazine nucleus, also shows PPAR activity and increases sensitivity to insulin as well. The synthesis of the side chain begins with Emmons–Smith condensation of benzaldehyde **16** with the ylide from phosphonate **17** and base to afford the enol ether **18** as a mixture of isomers. Hydrogenation of the resulting intermediate reduces the double bond and at the same time removes the benzyl group to afford the free phenol (**19**). Reaction of this compound with a hydrolase enzyme leads to selective hydrolysis of the ester, which leads to the (*S*) enantiomer (**20**). This kinetic resolution affords the acid as a virtually single stereoisomer. The carboxyl group is then protected for the next step as its isopropyl ester (**21**). In a convergent scheme, the anion on nitrogen from phenoxazine proper (**22**) and butyllithium is treated with ethylene oxide. The hydroxyl group on the end of the ethyl group in **23** is next converted to its mesylate (**24**). Displacement of this newly introduced leaving group by the phenoxide from **21** and potassium carbonate leads to the alkylated derivative **25**. Saponification of the terminal ester then yields **ragaglitazar** (**26**).[3]

Irritable bowel syndrome (IBS) is said to rank among the largest causes for physician office visits. There are, in spite of this, very few means for treating this very prevalent condition. The newly approved indication for the recently introduced 5-HT serotonin antagonist tegoserod (see Chapter 7) comprises one of the first specific treatments for IBS. Acylation of the indole (**27**) with trichloroacetyl chloride affords the ketone (**28**). Treatment of that intermediate with methanolic base leads to loss of the tri-chloromethyl group via the haloform reaction with concomitant esterification of the carbonyl group **29**. The benzyl protecting group is then replaced by acetyl with successive hydrogenation and reaction of the resulting phenol with acetic anhydride (**30**). Reaction of **30** with NCS leads to introduction of chlorine at the 2-position on the indole ring (**31**). This last intermediate is then treated with 3-chloropropan-1-ol in the presence of methylsulfonic acid, which forms the oxazine ring (**32**). Though the exact sequence is unstated, this reaction involves replacement of chlorine on the indole by oxygen and subsequent displacement of side-chain chlorine by indole nitrogen. The acetyl function on the phenol, having served its function, is next replaced by benzyl with successive saponification, and then alkylation of the resulting phenol with benzyl bromide (**33**). Heating **33** with the substituted piperidine (**34**), exchanges the methyl group on the ester with the primary amino group in **34** to form the corresponding amide (**35**). Catalytic hydrogenation then cleaves the benzyloxy group to yield the free phenol and thus the serotonin 5-HT receptor antagonist **piboserod** (**36**).[4]

DABCO = 1,4-Diazabicyclo[2,2,2]octane

The pituitary hormone arginine vasopressin (AVP) plays a pivotal role in regulating blood volume. Broadly speaking, excessive release of the hormone will cause the kidneys to reabsorb water and thus increase blood volume. The AVP antagonist would thus be useful in treating disease marked by excessive water retention, such as congestive heart failure. A pyrrolobenzazepine forms an important part of a non-peptide AVP antagonists. Acylation of the known pyrrolobenzazepine (**37**) with the nitrobenzoyl chloride **38** proceeds in straightforward fashion to the amide (**39**). Reduction of the nitro group by means of hydrogenation or stannous chloride affords the corresponding aniline (**40**). Acylation of the newly formed amino group with the benzoyl chloride (**41**) then affords **lixivaptan** (**42**). An alternate approach assembles the full side chain and then attaches that to the heterocycle **37**.[5]

Antiandrogens have proven very useful in treating BPH, an all too frequent accompaniment of aging. These agents, in addition, have a minor place in the treatment of prostatic cancer. They are also, on a more trivial note, used to reverse hair loss due to male pattern baldness. Drugs that act as antiandrogens, such as finasteride and dutasteride (Chapter 2), generally act by inhibiting the enzyme 5-α-reductase, which converts precursors, such as testosterone, to their active form by reducing the double bond at the 4-position in the steroid A ring. These drugs, formally derived from steroids, all retain the bulk of the precursor nucleus. A pair of more recent, closely related, compounds retain the lactam A ring of the steroid-derived agents, but replace steroid rings C and D by a simple aromatic ring. These agents retain the 5-α-reductase activity and have as a result been tested in the clinic as potential agents for treating cancer of the prostate. The first step in the synthesis of **bexlosteride** (**49**) comprises conversion of the tetralone (**43**) to its enamine by reaction with pyrrolidine. Reaction of **44** with a large excess of acrylamide under carefully controlled conditions leads to formation of the unsaturated lactam (**45**). The sequence can be rationalized by assuming that the first step comprises conjugate addition to the acrylamide; displacement of the pyrrolidine by amide nitrogen completes ring formation. The double bond at the ring fusion is next reduced with triethylsilane in the presence of trifluoroacetic acid. Thus, the saturated lactam that consists largely of racemic isomer with the trans ring fusion. The amide nitrogen in the product, which still contains a small amount of cis isomer, is next alkylated with methyl chloride in the presence of base to yield **46**. Reaction of that intermediate with methanol opens the lactam ring to yield the corresponding methyl ester **47**; the small amount of cis isomer can be separated at this stage since it resists methanolysis. The amino–ester is then resolved via its ditoluyl tartrate salt. Heating the resolved aminoester (**48**) with sodium carbonate then regenerates the lactam ring to afford **49**.[6]

The scheme used to produce a somewhat more complex 5-α-reductase inhibitor relies on a chiral auxiliary to yield the final product as a single enantiomer. The first step in a sequence similar to that above, starts with the reaction of bromotetralone (**50**) with (R)-α-phenethyl amine (**51**) to afford the enamine (**52**). Reaction with methyl iodide adds the methyl group at what will be a steroid-like AB ring junction. This product is then treated with acryloyl chloride. The initial step in this case probably involves acylation of nitrogen on the enamine; conjugate addition completes formation of the lactam ring. Treatment of **54** with triethylsilane then reduces the ring unsaturation and cleaves the benzylic nitrogen bond to yield **55** as the optically pure trans isomer. Displacement of bromine with the mercaptobenzthiasole (**56**) completes the synthesis of **izonsteride** (**57**).[7]

The aggregation of blood platelets, the prelude to formation of blood clots, is a very necessary process for preserving the integrity of the circulatory system. Inappropriate platelet aggregation on the other hand can result in the formation of clots that block vital organs leading to strokes and heart attacks. Various approaches have been followed in the search for agents that decrease platelet aggregability: Widely used clopidogrel, for example, blocks adenosine diphosphate receptors on platelets, newer candidates, such as the "gartans" (Chapter 1), act further down the line by inhibiting fibrinogen. The compound at hand **apafant (67)** antagonizes the action of the platelet activating factor (PAF), a substance that not only induces platelet aggregation, but is a factor in inflammatory processes and hypersensitivity. Reaction of the β-ketoaldehyde, diester **58** with the benzoyl nitrile (**59**) in the presence of sulfur leads to formation of the aminothiophene (**60**). Overlooking the diester for the moment, the reaction can be rationalized for bookkeeping purposes by assuming condensation of the aldehyde with the activated ketonitrile methylene group, conversion of the ketone to a thioketone and addition of that to the cyano group as its enolate, though not necessarily in that order. Heating the product in strong acid cause the diester in the product to decarboxylate. The resulting acid is then reesterified with methanol (**61**). Construction of the diazepine ring starts by alkylation of the amino group with bromoacetamide in the presence of base to afford **62**. Strong acid, for example, polyphosphoric acid, the causes the side-chain amide nitrogen to react with the ketone to form a cyclic imine affording **63**. The ester grouping is then restored and the amide carbonyl is converted to a thioamide with phosphorus pentoxide **64**. Reaction of this last intermediate with hydrazine gives the cyclic amino-amidine (**65**) putting in place the nitrogen atoms of the future triazole ring. Condensation of that product with ethylorthoacetate leads to formation of the last fused heterocyclic ring. Ester–amide interchange of this last product with morpholine completes the synthesis of **apafant** (**67**).[8]

The adventitious discovery of the utility of α-adrenergic blockers for treating benign prostatic hypetrophy led to the introduction of several of these agent. The majority of these compounds, for example, alfuzocin (Chapter 8), consist of modified quinazolines. A structurally quite distant heterocyclic compound has been found to have much the same activity at adrenergic receptors in experimental system. This activity extends to α-adrenergic receptors located in the prostate. One arm of the convergent synthesis starts with the condensation of the anion from dimethylresorcinol with DMF. Hydrolysis of the initial product then yields the aldehyde. Reaction with aluminum chloride leads to scission of one of the methyl ethers to give the phenol (**68**). Heating **68** with acetic anhydride probably leads initially to formation of the acetylated phenol. The presence of base then causes the phenol to cyclize to coumarin (**69**). Condensation of **69** with the azomethine ylide from **70** leads to 3 + 2 cycloadditon to the cou-marin double bond. The presence of the chiral auxiliary α-phenethylamine leads to the formation of the addition product as an essentially single enan-tiomer (**71**). Reduction of the coumarin carbonyl with lithium borohydride gives the ring-opened hydroxy phenol (**72**). A mixture of mesylates (phenol and hydroxymethyl) is obtained on treating **72** with methanesulfonyl chlor-ide. In the next step, treatment with strong base leads to internal displace-ment of the mesylate closing the ring to afford the corresponding benzopyran (**73**). Hydrogenation then cleaves the phenethyl group to afford product **74**, which contains the secondary amine required for coupling with the other major fragment.

Synthesis of the second heterocyclic fragment begins with the condensation of phenylglyoxal oxime (**75**) with guanidine to form the pyrazine N-oxide (**76**). Treatment with triethylphosphine reduces the N-oxide function leading to pyrazine (**77**). The amino group is next treated with nitrous acid and the resulting diazonium salt reacted with hydrogen bromide to afford the brominated derivative (**78**). Reaction of **78** with ethyl thioglycolate in the presence of sodium carbonate arguably begins by displacement of bromine by sulfur to form the transient thioether (**79**). Addition of the enolate on carbon adjacent to the ester then attacks the cyano group to form the fused thiophene ring **80**. The newly formed amine on the thiophene ring is then converted to the corresponding isocyanate (**81**), by reaction with phosgene.

There remains the task of joining together the two heterocyclic fragments. To this end the free secondary amino group on the benzoxazine (**74**) is alkylated with 3-bromopropionitrile; the terminal cyano group is then reduced to the primary amine by means of Raney nickel (**82**). Condensation of **82** with the pyridazothiophene (**81**) leads to addition of the amine to the isocyano group to afford the transient coupled urea (**83**). The urea nitrogen furthest from the ring then displaces the ethoxide from the nearby ester group to form a fused pyrimidone ring. Thus, the α_1-adrenergic blocker **fiduxosin** (**84**) is finally obtained.[9]

Anthraquinones have been investigated in some detail over the past several decades as sources for antitumor agents. Several compounds whose structure includes this three-ring fragment have gone as far as the clinic. It has since been established that cytotoxic activity of these compounds is due mainly to their activity on topoisomerase 2. The anthraquinone mitoxantrone, was approved by the FDA close to two decades ago. An analogue in which one of the benzene rings is replaced by pyridine, perhaps not surprisingly, retains antitumor activity. Reaction of the lithio reagent from 4-chlorofluorobenzene (**86**) with the pyridine equivalent (**85**) of phthalic anhydride affords the acylation product (**87**). Treatment with acid leads to internal acylation and formation of the aza-anthraquinone (**88**). Condensation of this intermediate with the substituted hydrazine (**89**) leads to formation of the fused pyrrazole (**90**). The regiochemistry of this reaction would suggest that the first step involves displacement of halogen and is thus guided by greater ease of replacing fluorine over chlorine; ordinary imine formation then closes the ring. Displacement of chlorine by N,N-dimethylethylenediamine (**91**) completes the synthesis of **topixantrone** (**92**).[10]

2. COMPOUNDS WITH FOUR FUSED RINGS

The serious side effects associated with long-term use of dopamine antagonist antipsychotic agents have led to the continuing search for better
tolerated drugs. This research has to some extent been encouraged by
the identification of ever more subtypes of dopamine receptors; this
raises the possibility there may be drugs found that act on a subtype receptor more specific for treating schizophrenia than for the receptors associated with side effects. **Ecopipam (101)** which differs markedly in
structure from most of the widely used antipsychotic drugs, is specific
for D1 dopamine receptors. Though this compound was found to be not
particularly effective for treating schizoid patients, it does have an effect
against addictive behaviors. One of several routes to this compound
starts with the decalin trans aminoalcohol (**93**). This compound is then
alkylated with the methyl acetal (**94**) from bromoacetaldehyde. The ring
hydroxyl in **95** is then converted to its chlorophenylsulfonate (**96**). This
intermediate spontaneously closes to the aziridinium salt (**97**) by displacement of the sulfonate by the adjacent amine. Condensation of **97** with
Grignard reagent **98** leads to addition at the tertralin 1 position with concomitant ring opening to yield **99**. Treatment of **99** with strong acid
leads to attack of the newly revealed aldehyde on the aromatic ring and
thus formation of the azepine ring. The double bond formed in the
course of this last reaction is then reduced with diborane (**100**). Scission
of the methyl ether with boron trichloride the affords (**101**).[11]

Yet another potential antipsychotic drug with an unusual pattern of
receptor affinity consists of a dibenzoxapine. This compound **asenapine**

(**107**) was interestingly first described in a 1979 patent.[12] Data on this agent using more recent pharmacological methods apparently led to its being pulled off the shelf. Condensation of the acid chloride (**102**) with N-methylglycine ethyl ester leads to the amide (**103**). Treatment of **103** with potassium *tert*-butoxide leads the enolate adjacent to the aromatic ring to add to the ester at the end of the side chain, thus forming the pyrrolidine ring (**104 = 105**). Heating this intermediate (**105**) in PPA leads to reaction of the ketone carbonyl group with the other aromatic ring to form the benzoxazine (**106**). The unsaturation in the seven-membered ring is then reduced by means of sodium in liquid ammonia leading to the product with a trans ring junction. The amide function is then taken on the amine by means of a metal hydride. Resolution, not described in the patent, will then afford chiral **107**.

The structures of the vast majority of PD-5 inhibitor compounds aimed at erectile dysfunction consist of modified purines. The structure of the recently approved drug for this indication **tadalafil** (**113**) differs markedly from the prototypes. Tryptophan methyl ester (**108**) provides the starting material for large scale enantioselective synthesis. Condensation of that compound with piperonal (**109**) in the presence of acid leads to formation of the tricyclic intermediate (**110**). This transform involves initial addition of the amine to the aldehyde. The carbocation from the newly formed carbinolamine then attacks the indole 2-position to form the the fused piperidine. The stereochemistry of the new chiral center is guided by that from the tryptophan carbon across the ring. The secondary amine is next acylated with chloroacetyl chloride in the presence of triethylamine to afford **111**. Reaction of this intermediate with methylamine goes on to form the desired product in a single step. This reaction can be rationalized by assuming initial displacement of terminal chlorine by the amine to give the transient intermediate **112**. This amino group then takes part in an ester–amide interchange in the presence of base to form the new ring. Thus, **113** is obtained.[13]

3. COMPOUNDS WITH FIVE OR MORE FUSED RINGS: CAMPTOTHECINS

Though the potent cytotoxic activity of camptothecin (**114**) was recognized by the middle 1960s, the drug was not used in the clinic until just over a decade ago. The very poor solubility of the compound initially led to the use in clinics of the ring opened acid. It was later found that the intact lactone was essential for activity. Clinical trials in the early 1990s confirmed that finding. The mechanism of action of this drug as an inhibitor of topoisomerase I sets it aside from many other widely used compounds, such as the anthracyclines that act on topoisomerase II. A number of analogues that mainly include the addition of solubilizing functions, such as topotecan, lurtotecan, and irinotecan, and have been approved for use in patients. An intermediate on the way to a compound that includes a solubilizing amino group has interestingly been assigned a nonproprietary name. This compound **camptogen (115)** is available from the natural product in a single step using classical nitration conditions.[14]

The production by total synthesis of an analogue that incorporates an additional fused ring, by way of contrast, includes a good many more steps. The lengthy synthesis starts with the aluminum chloride-catalyzed acylation of fluorotoluene (**116**) with succinic anhydride to afford the acid (**117**). The ketone group is then reduced to methylene by means of hydrogen over palladium; the acid is then esterified with methanol to yield **118**. Reaction with nitric acid proceeds in a straightforward manner to afford the nitro derivative (**119**). The ester grouping is then saponified. Heating the resulting acid in polyphosphoric acid results in ring closure and thus formation of a tetralone ring (**120**). The next few steps establish the requisite functionality in this newly added ring. This sequence includes formal transfer of the carbonyl group to the alternate benzylic position.

Sequential reduction of the ketone and dehydration of the resulting alcohol introduces a double bond. Hydrogenation then both reduces the unsaturation and takes the nitro group on to the amine (**121**). The amino group is then acylated with acetic anhydride. Reaction with permanganate introduces a ketone at the alternate position from that remaining after cyclization (**122**). Reaction of this last intermediate with butyl nitrite in the presence of strong base introduces nitrogen at the position adjacent to the ketone in the form of an oxime (**123**). This group is then reduced to the corresponding amine by means of zinc in acetic acid and anhydride. This reaction affords the acetamide (**124**). Sequential hydrolysis of the amides followed by acylation with trifluorcacetyl chloride yields the amide (**125**). This last reaction is apparently specific for the amine adjacent to the ketone, leaving that on the aromatic ring as a free amine. This bicyclic compound is then condensed with the known total synthesis intermediate **126**, in the presence of acid to afford **exatecan** (**127**).[15] Again, for bookkeeping purposes, this last transform can be viewed as reaction of the ketone in **126** with the aniline nitrogen and condensation of the resulting enamine with the tetralone carbonyl group.

REFERENCES

1. D. Cook, J.E. Meritt, A. Nerio, H. Rapoport, A. Stassinopoulos, S. Wolowitz, J. Matejovic, W.A. Denny, U.S. Patent 6,514,987 (2003).
2. J.E. Wrobel, A.J. Dietrich, M.M. Antane, U.S. Patent 6,251,936 (2001).
3. S. Ebdrup et al., *J. Med. Chem.* **46**, 1306 (2003).
4. M. Fedouloff, F. Hossner, M. Voyle, J. Ranson, J. Powles, G. Riley, G. Sanger, *Bioorg. Med. Chem.* **9**, 2119 (2001).
5. J.D. Albright et al., *J. Med. Chem.* **41**, 2442 (1998).
6. B.A. Astelford, J.E. Audia, J. Deeter, P.C. Heath, S.K. Janisse, T.J. Kress, J.P. Wepsiec, L.O. Weigel, *J. Org. Chem.* **61**, 4450 (1996).
7. J.R. Audia, L.A. McQuaid, B.L. Neubauer, V.P. Rocco, U.S. Patent 5,662,962 (1997).
8. K.H. Weber, *Drugs Future* **13**, 242 (1988).
9. A.R. Haight et al., *Org. Proc. Res. Devel.* **8**, 897 (2004).
10. A.P. Krapcho, E. Menta, A. Oliva, S. Spinelli, U.S. Patent 5,596,097 (1997).
11. D. Hou, D. Schumacher, *Curr. Opinion Drug Res. Devel.* **4**, 792 (2001).
12. J. van der Burg, U.S. Patent 4,145,434.
13. P.J. Dunn, *Org. Proc. Res. Devel.* **9**, 88 (2005).
14. M. Wani, A.W. Nicholas, M.E. Wall, *J. Med. Chem.* **29**, 2358 (1986).
15. H. Terasawa, A. Ejima, S. Ohsuki, K. Uoto, U.S. Patent 5,834,476 (1998).

SUBJECT INDEX

Bold numerals refer to syntheses

The Organic Chemistry of Drug Synthesis, Volume 7. By Daniel Lednicer
Copyright © 2008 John Wiley & Sons, Inc.

CROSS INDEX OF BIOLOGICAL ACTIVITY

The Organic Chemistry of Drug Synthesis, Volume 7. By Daniel Lednicer
Copyright © 2008 John Wiley & Sons, Inc.

Antibacterial (*Continued*)
Sitafloxacin, 174
Antidepressant
Citalopram, 140
Elzasonan, 136
Flibanserin, 158
Igmesine, 64
Lubazodone, 70
Reboxetine, 61
Vofopitant, 110
Antidiabetic
Ertiprotafib, 219
Muraglitazar, 47
Nateglinide, 15
Netoglitazone, 103
Ragaglitazar, 220
Repaglinide, 15
Rivoglitazone, 157
Sitagliptan, 206
Vidagliptin, 84
Antiemetic
Aprepitant, 105
Ezlopitant, 62
Maropitant, 62
Antiepileptic
Rufinamide, 108
Antiestrogen
Acolbifene, 163
Arzoxyfene, 155
Bazedoxifene, 145
Fulvestrant, 32
Lasofoxifene, 77
Ospemifene, 65
Antifungal
Omoconazole, 94
Posaconazole, 105
Voriconazole, 103
Antihpertensive
Atrasentan, 87
Gemopatrilat, 19
Clevipidine, 117
Edonentan, 56
Nepicastat, 97
Olmesartan, 111
Antiinflammatory
Darbufelone, 99
Ecopladib, 147
Pelitinib, 167

Varespladib, 144
Antiinflammatory COX-2
Cimicoxib, 96
Etoricoxib, 117
Parecoxib, 92
Tilmacoxib, 91
Valdecoxib, 92
Antiparasitic
Nitazoxinide, 101
Antiprogestin
Asoprinosil, 32
Antipsychotic
Asenapine, 228
Bifeprunox, 159
Ecopipam, 228
Paliperidone, 209
Sonepiprazole, 166
Antitumor
Alitretoin, 30
Alvocidib, 165
Apaziquone, 154
Bexarotene, 76
Bortezomib, 133
Camptogen, 230
Canertinib, 180
Canfosfamide, 13
Clofarabine, 200
Cloretazine, 14
Doramapimod, 74
Emitefur, 127
Enzastaurin, 89
Erlotinib, 179
Exatecan, 232
Forodesine, 192
Gefitinib, 181
Irofulven, 38
Istradefylline, 202
Lapatinib, 182
Monasterol, 128
Mubritinib, 108
Nelarabine, 200
Pelitrexol, 212
Pemetrexed, 193
Pralatrexate, 211
Prinomastat, 54
Semaxanib, 149
Sunitinib, 150
Talabostat, 84

CUMULATIVE INDEX